Lisbeth Borbye, Michael Stocum,
Alan Woodall, Cedric Pearce,
Elaine Sale, William Barrett,
Lucia Clontz, Amy Peterson,
and John Shaeffer

Industry Immersion Learning

Related Titles

Behme, S.

Manufacturing of Pharmaceutical Proteins

2009
ISBN: 978-3-527-32444-6

Gruber, A. C.

Biotech Funding Trends
Insights from Entrepreneurs and Investors

2009
ISBN: 978-3-527-32435-4

Meibohm, B. (ed.)

Pharmacokinetics and Pharmacodynamics of Biotech Drugs
Principles and Case Studies in Drug Development

2006
ISBN: 978-3-527-31408-9

Kerzner, H.

Project Management Case Studies

2005
ISBN: 978-0-471-75167-0

Lisbeth Borbye, Michael Stocum, Alan Woodall,
Cedric Pearce, Elaine Sale, William Barrett, Lucia Clontz,
Amy Peterson, and John Shaeffer

Industry Immersion Learning

Real-Life Industry Case-Studies
in Biotechnology and Business

**WILEY-
BLACKWELL**

WILEY-VCH Verlag GmbH & Co. KGaA

The Authors

Dr. Lisbeth Borbye

Michael Stocum

Alan Woodall

Cedric Pearce

Elaine Sale

William Barrett

Lucia Clontz

Amy Peterson

John Shaeffer

■ All books published by Wiley-VCH are carefully produced. Nevertheless, authors, editors, and publisher do not warrant the information contained in these books, including this book, to be free of errors. Readers are advised to keep in mind that statements, data, illustrations, procedural details or other items may inadvertently be inaccurate.

Library of Congress Card No.: applied for

British Library Cataloguing-in-Publication Data
A catalogue record for this book is available from the British Library

Bibliographic information published by the Deutsche Nationalbibliothek
The Deutsche Nationalbibliothek lists this publication in the Deutsche Nationalbibliografie; detailed bibliographic data are available on the Internet at http://dnb.d-nb.de

© 2009 WILEY-VCH Verlag GmbH & Co. KGaA, Weinheim, Germany

Composition K+V Fotosatz GmbH, Beerfelden
Printing betz-druck GmbH, Darmstadt
Bookbinding Litges & Dopf GmbH, Heppenheim

Printed in the Federal Republic of Germany
Printed on acid-free paper

ISBN 978-3-527-32408-8

Contents

Industry Immersion Learning. Real-Life Industry Case-Studies in Biotechnology and Business
L. Borbye, M. Stocum, A. Woodall, C. Pearce, E. Sale, W. Barrett, L. Clontz, A. Peterson, J. Shaeffer
Copyright © 2009 WILEY-VCH Verlag GmbH & Co. KGaA, Weinheim
ISBN: 978-3-527-32408-8

Preface

Graduates who secure their first jobs in industry typically spend a significant amount of time adjusting to the new environment because it is so different from the traditional university setting. Together with multiple industry professionals in the Research Triangle Park, North Carolina, I have introduced a learning method called "industry immersion learning" with the goal of easing the transition from the university to the workplace (here, the biotechnology industry).

The industry immersion method is characterized by need-based, innovation-oriented, and proactive acquisition of knowledge. The education is coordinated and supervised by academic and industry professionals in concert and promotes a high level of interaction between students and industry professionals. As the name of the method implies, students are immersed in the industry environment and tasked to excel in matters of high relevance to the company in which the training takes place. Students must adapt quickly to the new environment, create a professional network on site, become knowledgeable about the topic of study, employ innovative thinking, and meet or exceed expectations in their deliverables in a timely manner in order to succeed.

The industry immersion method has been received with enthusiasm among students and both university and industry leaders. It provides a means for the students to graduate with an industry-relevant education, and the university to provide industry with a better prepared, industry-ready workforce while simultaneously creating important university–industry networks and empowering employers to participate in curriculum design.

In an attempt to disseminate the method to a larger audience, employer alliance building and the industry immersion method are described in detail in this book, and seven industry projects, the so-called "case studies", have been compiled and transposed to a format useful in both industry and classroom settings. Each of these sample industry case studies focuses on a particular trend and together they provide a nonexhaustive view into selected, timely topics. The logistics of teaching by immersion are outlined and a variety of parameters can be customized to match the environment in which they are taught. By consolidating these examples of industry case studies in this book, I encourage their "re-use" while simultaneously hoping to inspire the creation of many new case studies and much new collaboration between universities and industry.

Industry Immersion Learning. Real-Life Industry Case-Studies in Biotechnology and Business
L. Borbye, M. Stocum, A. Woodall, C. Pearce, E. Sale, W. Barrett, L. Clontz, A. Peterson, J. Shaeffer
Copyright © 2009 WILEY-VCH Verlag GmbH & Co. KGaA, Weinheim
ISBN: 978-3-527-32408-8

I am greatly indebted to the many industry professionals who have volunteered their time, expertise, effort, and enthusiasm to help me establish industry immersion learning at North Carolina State University. I am also grateful to the many students who bravely embraced the industry case studies and industry immersion education, displayed immense personal and professional growth, and commenced exciting careers with a skills set aligned with employers' needs.

The generosity of the North Carolina Biotechnology Center made this publication possible.

Raleigh, December 2008   *L. Borbye*

Disclaimer

The content of this book is based on the individual authors' personal knowledge and experiences. This book presents selected topics in the biotechnology and pharmaceutical industries. Mentioned laws, regulations, and guidelines are based on current status at the time of the authors' experience or writing. Each author is solely responsible for the content of his or her own chapter. The authors and editor disclaim any liability or loss in connection with use of the information herein. Use of the information is at own risk.

The contents of this publication and related links do not constitute legal or professional advice. Readers should not act or rely on any information in this book without first seeking the advice of an attorney or other relevant expert.

No part of this publication and the web presentations may be reproduced, stored in a retrieval system, or transmitted, in any form or by any means, electronic, mechanical, photocopying, recording, or otherwise, without the prior written permission of the publisher and relevant author(s).

Industry Immersion Learning. Real-Life Industry Case-Studies in Biotechnology and Business
L. Borbye, M. Stocum, A. Woodall, C. Pearce, E. Sale, W. Barrett, L. Clontz, A. Peterson, J. Shaeffer
Copyright © 2009 WILEY-VCH Verlag GmbH & Co. KGaA, Weinheim
ISBN: 978-3-527-32408-8

1
Principles of Industry Immersion Learning

Lisbeth Borbye

Contents

Industry Immersion Learning. Real-Life Industry Case-Studies in Biotechnology and Business
L. Borbye, M. Stocum, A. Woodall, C. Pearce, E. Sale, W. Barrett, L. Clontz, A. Peterson, J. Shaeffer
Copyright © 2009 WILEY-VCH Verlag GmbH & Co. KGaA, Weinheim
ISBN: 978-3-527-32408-8

Abbreviations

GMP = Good Manufacturing Practice
cGMP = current Good Manufacturing Practice
GLP = Good Laboratory Practice
CEO = Chief Executive Officer
FDA = Food and Drug Administration
SWOT = Strengths, Weaknesses, Opportunities, Threats

1.1
Introduction

Traditionally, universities have produced the same kind of employees for both academic and industry work environments. The industry work environment has changed dramatically during the last two decades and the skill set needed in industry today is very different from the one needed in academia. It includes a high level of technical aptitude, multiple professional competencies, an interdisciplinary, highly flexible, and collaborative attitude, and a globally oriented perspective.

Coming from a traditional university training, graduating students face a highly challenging work environment when they enter industry careers. The university education is typically acquired through content-oriented classroom lectures and hands-on laboratory work. It promotes the students' analytical and individual skill sets and their ability to compete. Students gain a sharply defined amount of understanding in discrete topics, often in a nonintegrative manner. Industry needs a workforce with skills that both include the academic background and extend it. Prospective employees need to learn about industry-relevant topics, to understand and be able to operate in a context-oriented manner, to think innovatively, and to develop and utilize good communication and interpersonal skills through teamwork and networking.

As a response to this need, universities in the United States and elsewhere are showing interest in need-based curricula and a concept called "professional Master's education". The goal is to tailor professional graduate education to meet employers' needs. These degree programs focus on developing employer-relevant education, primarily by including new topics and often multi- and/or interdisciplinary training in their curricula. The programs vary in their levels of interaction with industry. An example of this type of program which employs multifaceted interaction with many industry professionals in the Research Triangle Park, North Carolina, is the professional Master's program in Microbial Biotechnology at North Carolina State University. This program integrates academic and professional training in both business and science. Students learn work-force-related skills through industry internships, via industry mentors, and in a new course entitled "Industry Case Studies". This course is interdisciplinary and encompasses a variety of business and science initiatives. The Industry Case Studies course serendipitously utilizes components of action-based learning and context- and problem-based learning that involves cutting-edge unresolved projects and teamwork. The case studies are often "open-ended" which means there is a certain amount of flexibility concerning the topic and the outcome. This course is currently the only one of its kind; it employs so-called "industry immersion learning" and is tailored specifically to the biotechnology and pharmaceutical work environments in the Research Triangle Park. The following sections describe how this course was developed. Sections 1.2 and 1.3 have been reprinted with permission from *Journal of the Academy of Business education*.

1.2
Building a University – Industry Alliance

1.2.1
Educational Needs Assessment

In order to create employer-relevant education it is essential to become knowledgeable about the employers' needs. An effective way to identify immediate training needs is to survey professional employers regarding which skills graduates should have in order to obtain employment. Another method is the creation of an inventory map of employers, their fields of expertise, technology, market, size, and predicted growth in a particular area. Using a compilation of these methods, a list of highly desirable skills for graduate-level employees has been assembled (Table 1.1). Many of these represent a large challenge for universities, and effective training in these areas often requires a high level of interaction with professionals. This interaction can take many forms including internships, guest instruction on projects/case studies, guest lectures, and mentoring.

1.2.2
Establishing Contact

After the determination of educational needs and resources, key contributors from industry must be identified. This process involves research of individual industry professionals, their positions and responsibilities, fields of expertise, and level of interest and skills in curriculum design and teaching. Phone calls and visits must be made and often followed up with numerous meetings to discuss the scope and length of the interaction, necessary resources, and format. The interaction must be initiated at the correct level. The industry professional should have the flexibility to make the decision to interact with academia without jeopardizing company interests. As an example, a request for interaction is more likely to receive a response from the chief executive officer (CEO) of a small (less than 50 employees) or medium-sized company (50–500 employees) than from the CEO of a large company (more than 500 employees). This is a reflection of responsibilities and resource constraints. Table 1.2 lists examples of key contributing industry professionals according to their position and the size of the company in which they are employed.

Table 1.1 Skills in high demand in the biotechnology and pharmaceutical industries.

Competency category	Discipline
1. Academic	Science
	Business
	Integration of science and business
	Analytical thinking
2. Practical experience	Internship
	Bench and office work
3. Technological	Research and development process
	Manufacturing process
	Clinical trials process
4. Specialty	Intellectual property
	Regulatory knowledge
	GMP/GLP exposure
	Project management
	Accounting and introductory
	Finance
5. Soft skills	Leadership
	Mentorship
	Teambuilding
	Conflict management
	Expectation management
	Change management
	Ambiguity management
	Communication skills
6. Mindset	Context-oriented thinking
	Out-of-the-box focus
	Entrepreneurship
	Global orientation

GMP, good manufacturing practice; GLP, good laboratory practice

Table 1.2 Industry professionals: rank of contact person and company size.

Contact person	Company size
Team leader	Large
Manager	Large
Director	Large
Vice president	Large
Director	Medium
Vice president	Medium
Vice president	Small
CEO	Small

1.2.3
Marketing Incentives

Clearly describing the advantages of collaborative teaching is essential for alliance building. As examples, industry professionals have an opportunity to:

- obtain resources through student work and new ideas;
- train students in skills of importance for industry;
- evaluate students for future employment;
- gain access to potential employees.

Academics, in return, have the potential to access new technology, interdisciplinary thinking, and know-how, as well as to gain a professional network by interacting with industry professionals.

1.2.4
Obtaining Commitment

After initial interest has been spurred, reciprocal goals must be set. Flexibility should be displayed in any or all of the following areas: timing, length of project, participating instructors, specific deliverables (such as reports and presentations), intellectual property matters, and geographical logistics. Intellectual property is often a matter of great concern for both parties. Universities typically want to protect students and faculty from signing unnecessary contracts of restraint and desire to limit the length of the contract. Industry professionals, on the contrary, have a duty to protect the intellectual property and trade secrets owned by the company of employment. Agreements that provide for nonconfidential interaction are easier to manage, but the outcomes are limited by disclosure restrictions (see also 1.8.7 Legal Requirements).

1.2.5
Alliance Dynamics

Many factors influence the dynamics of the academia–industry interaction. If the interaction is deemed positive, it is likely to continue and become an asset for both parties and an opportunity for further development. External factors play an important role and can cause unexpected change with demands on flexibility. Examples might include sudden travel arrangements, unforeseen audits, termination of employment, relocation, company reorganization, altered priorities caused by market forces, lack of resources, promotions, mergers and acquisitions, and personal reasons. Success and maintenance of an interaction between industry professionals and academics are subject to such changes, and the ability to quickly adjust to alternative strategies when change happens is essential.

1.3
Design, Format, and Model Examples of Case Studies

The broad scope of intellectual exchange defined by alliances becomes a useful platform for innovation. Due to the synergistic nature of the interaction, new education not normally employed at universities is being created; an example is the Industry Case Studies course. Industry professionals and faculty discuss matters concomitantly from both an industry and a university perspective. They decide topics, process and teaching methods, duration, geographical location, re-source allocation, deliverables, and target audience for final presentations. They discuss and resolve legal matters and may agree to publicize the interaction with academia and with the students through press releases or other media. Table 1.3 displays a summary of case studies as they were created and per-formed in the period from August 2003 (program launch) to May 2008. All case studies target the core competencies listed in Table 1.1 and involve the study of a specific forefront issue in science and/or business.

1.3.1
Example 1: Technology Development

Case study #5 focused on the manufacture of enzymes and was performed at a large biotechnology company. Students learned about the process of enzyme

Table 1.3 Examples of case studies and timelines.

Case study	Topic	Duration
1	Gene expression	3 weeks
2	High throughput diagnostics	4 weeks
3	Modified crops	4 weeks
4	Bioremediation	4 weeks
5	Technology development	4 weeks
6	Quality assurance, cGMP training	7 weeks
7	Biomarkers *(Chapter 2)*	6 weeks
8	Communication skills, conflict management	2 weeks
9	Patent law *(Chapter 5)*	8 weeks
10	Product assessment	4 weeks
11	FDA audit	2 weeks
12	Business development	11 weeks
13	Project management *(Chapter 3)*	8 weeks
14	Human error prevention *(Chapter 8)*	8 weeks
15	Intellectual property management *(Chapter 6)*	8 weeks
16	Outsourcing vs. in-house technology	6 weeks
17	Entrepreneurship *(Chapter 4)*	8 weeks
18	Operational excellence in manufacturing *(Chapter 7)*	8 weeks
19	Real-time monitoring of contamination and resources	12 weeks
20	Minimizing company environmental footprint	12 weeks

cGMP, current good manufacturing practice; FDA, Food and Drug Administration.

production and were tasked to create methods to minimize fouling on filtration membranes by microorganisms (the occurrence of so-called "biofilm"). Students worked to comprehend a variety of issues in the areas of microbiology, biochemistry, molecular biology, and engineering. They suggested a mix of physical, chemical, and biological barriers as a means to reduce biofilms. The students finalized the project with a report and a presentation for the employees at the plant. The case study gave the students an opportunity to work as entrepreneurs on an important problem in industry, work interdisciplinarily, practice teamwork, improve their oral and written communication skills, and present their ideas to a large heterogeneous group of industry professionals including laboratory staff members and executive officers.

1.3.2
Example 2: Product Assessment

Case study #10 took place in a medium-sized biotechnology company. The project involved analysis of a confidential new product. Students were divided into two teams, a business team and a science team. The business team was tasked to perform a market analysis, investigate the intellectual property status, develop a budget, and predict the profit margin. The science team developed protocols, proposed a timeline for production, assessed the development costs, and considered compliance issues. Together the teams developed a business plan and gave their recommendations. Students developed an understanding of the challenges of science and business teams working together, components of a business plan, assessing the viability of a product, justifying the assessment to an executive management team, and maintaining professional conduct while working on a confidential project.

1.3.3
Example 3: Business Development

The goal of case study #12 was innovation and marketing of new products containing antioxidants. This study was carried out in a large pharmaceutical company. The students performed a cost–benefit analysis and a market analysis that included identifying the current competition. They studied a variety of antioxidants and compiled and summarized large amounts of data on the biochemical pathways involved including toxicological and pharmacokinetic aspects. The students decided to present to the management team product ideas useful for both prophylaxis and treatment of selected conditions. During the process many product ideas were deemed not profitable and subsequently rejected. This study gave the students an opportunity to comprehend the amount of business and science information that is necessary to analyze a new product idea, learn that projects are often terminated and that flexibility is a necessary skill, design a new product, market their product idea in a corporate forum, and discuss their product idea with a professional management team.

1.4
Basics of Industry Immersion Learning

1.4.1
Definition and Characterization

Immersion learning is defined here as the learning that occurs as a result of immersion in a particular environment. Therefore, it is characterized as an environment-related and environment-preparative method that employs aspects of multiple learning methods in addition to its unique environmentally dependent features.

1.4.2
The Immersion Environment

Before immersion learning can be initiated, industry professionals and academics must have agreed to and prepared projects within the relevant frameworks for the students. Because the projects typically are open-ended, there is no result indication beforehand, and often projects address high-risk areas in need of investigation and resources. Consequently, students are placed directly in the relevant environment, which may be unfamiliar, or slightly familiar, or even familiar to them. Because work environments and companies vary, no learning obtained in one environment will be absolutely replicable in another. This adds to the potential for understanding context.

1.4.3
Sample Work Flow of an Immersion Case Study

Teams of students are given rudimentary but necessary precursory information and training to present goals for the project, a plan for their activities, and a list of the deliverables (e.g., reports and presentations) they will provide. This plan must be accepted by the industry professionals, after which the students will be left to pursue their plan. This often requires extensive information exchange and building and creating of discussion networks on- and off-site as well as adjusting project goals and terms. The students finish the project by presenting their recommendations to the upper management in charge of the project and accepting feedback, questions, and suggestions for new deliverables.

1.4.4
Interactive Agents

One of the advantages of the immersion learning method is the need to find information that is not readily available in traditional formats (databases, textbooks, libraries, etc.). Students must seek advice and information from many sources, such as industry professionals, academics, vendors, catalogues, patent databases, to mention a few examples. They learn to identify and create alliances with these parties in order to achieve their goals, deadlines, and deliverables. In addition, students are coached by managers of the projects as well as the academics overseeing the immersion education, which create opportunities for understanding aspects from the viewpoint of both environments.

1.5
Predicted Learning Outcomes

There are four primary categories of immersion learning. These are: (1) specialty knowledge related to the environment, (2) professional, compliant conduct related to the environment, (3) interpersonal, communication, and networking skills, and (4) entrepreneurial mindset. The extent and type of learning can vary depending on how familiar the students are with the topic and the environment before immersion.

It is expected that students will adjust to the immersion environment and adapt attitudes and behaviors compliant with the particular environment's guidelines. Students are expected to learn by "living" in the environment and by permitting "knowledge diffusion" from the environment. The learning is expected to be context-oriented and related to topics and conduct relevant for the environment.

Because industry immersion fosters extraversion, interaction, teamwork, and networking, students are expected to develop their communication, and interpersonal skills as well as to become comfortable working in teams. It is expected that students learn to interact well with others, to share their ideas, and to contribute in a consistent manner to group goals while maintaining personal integrity, responsibility, and a professional identity. In addition, students are learning to manage unforeseen challenges relating to the specific environment. These can take the form of changes in a variety of parameters such as group composition, topic matter, deadlines, and instructor availability.

Innovative thinking constitutes another expectation. Students are stimulated to solve as yet unsolved problems with real-life implications. Most students find this very intriguing. They must decide how to proceed and how and where to retrieve information to make valuable recommendations. Because both team and personal reputations (including academic grades and later job opportunities) will be influenced by the quality of the work the students produce, they have strong incentives to excel.

1.6
Assessment of Actual Learning Outcomes

Due to the very complex nature of industry immersion learning, it can be an elaborate task to measure its outcome. Understanding additional learning parameters can be achieved by observing the actual challenges the immersion teaching method causes during the teaching period. In addition, students may reflect on their learning in a long range of areas such as the level of acquisition of science and business knowledge, presentation skills, interpersonal skills, flexibility, discipline, ambiguity management, cross-cultural understanding, and entrepreneurship.

1.7
Overview of Selected Case Studies

This book contains seven examples of educational industry case studies contributed by experts in various areas of biotechnology. These are by no means exhaustive of the field or an attempt to mimic a textbook; rather, these industry case studies represent highly selective subject matter for advanced students. The case studies include important topics such as:

- An overview of diagnostics technology and the challenges in drug discovery linking the identification and treatment of certain conditions
- Important considerations that arise when companies decide how to prioritize certain conditions and treatments
- Analysis of new business ideas and how to start new companies based on these ideas
- How to become knowledgeable about intellectual property and how to protect it
- How to organize and manage intellectual property and obtain an overview of the intellectual property landscape
- Introduction to the concept of operational excellence through systematic improvement of technology and processes
- Understanding how humans cause errors and how companies can attempt to remove the root causes of such errors

The first case study describes *personalized medicine* and is contributed by Michael Stocum. This chapter provides learning about the development of personalized medicines, drug discovery, biomarkers, breast cancer, and a breast cancer drug tailored to a specific segment of the population. This case study was taught during spring 2004.

Alan Woodall is the contributor of the second case study, a study of *drug portfolio management* with a focus on drug development choice and prioritization from the company perspective. This case study was taught during spring 2005 and 2007.

The third case study, in *entrepreneurship*, provides an outline of what is necessary to start a company. Cedric Pearce contributes this case study, which involves many important phases of entrepreneurship, such as conceiving and analyzing a business idea and writing a business plan. This case study was taught during spring 2006 and 2007.

Understanding of patent law is essential for all companies. Elaine Sale contributes the fourth case study, which provides basic knowledge about this topic. This case study was taught during fall 2004.

The *management of intellectual property* and how to develop a strategic and well-balanced overview of the competition and prospects are the subjects of the fifth case study, provided by Bill Barrett. This case study was taught during spring 2006.

Lucia Clontz contributes the sixth case study, which consists of two sub case studies in the area of *operational excellence*. She addresses both process optimization and technology improvement. This case study was taught during fall 2006.

Finally, the seventh case study addresses how humans can be taught to make fewer mistakes. Amy Peterson and John Shaeffer both work in the area of *job observation*, a discipline that focuses on preventing errors rather than reacting to them after they have occurred. This case study was taught during fall 2005.

1.8
Logistics of Industry Immersion Teaching

1.8.1
Topic Selection

When choosing a subject of study, it is essential that the topic is timely and of high priority in industry. Most often, a topic lends itself to additional and related topics because immersion learning is context-based. An instructor in industry immersion education will quickly realize that there are many opportunities for teaching and learning beyond the subject matter. I will illustrate this fact by using the industry case studies in this book as examples; Table 1.4 lists the subjects of study and the anticipated areas of learning together with some of the extended teaching opportunities.

1.8.2
Instructor and Instructor Affiliation

Industry topics are best taught by experts in the particular field. A willing expert is an individual who is knowledgeable about the field and abreast of issues relating to the topic, who has extensive experience, and who is available to perform or assist with teaching. It is wise to select instructors affiliated with companies that display sincere interest in the topic through relevant research and development efforts.

Table 1.4 Case study topics, anticipated learning and additional teaching opportunities.

Case study topic	Anticipated learning	Examples of extended teaching opportunities
Biomarkers	Diagnostics Drug discovery Co-development of drugs and diagnostics Translational research Breast cancer Personalized medicine Generic product profile	Genomics Molecular biology Cancer SWOT analysis
Drug portfolio management	Financial assessment concepts Financial assessment methods Target product profile Drug discovery Alzheimer's disease Anxiety Analgesia Cognitive enhancement Nicotine addiction	FDA regulations Marketing Communication Elevator speech Market analysis
Entrepreneurship	Entrepreneur characteristics Business plan Company organization Analysis of new business idea Ways to finance a business venture Innovation	Market analysis Grant writing Elevator speech Public presentations Marketing
Patent law	United States patent law Claims language Understanding inventions Protection of intellectual property Basic intellectual property transactions	Genomics Molecular biology Drug discovery FDA regulations Intellectual property and globalization Erythropoietin Biotechnology
Intellectual property management	Patent searching Patent mapping Patent claiming strategy	Genomics Drug discovery FDA regulations Biotechnology Presentation of large data sets
Operational excellence	Drug manufacturing Operational excellence Standards in manufacturing FDA regulations Six Sigma Lean manufacturing Biofilms	Drug manufacturing (detailed) Technology in the pharmaceutical industry Microbiology Engineering

Table 1.4 (continued)

Case study topic	Anticipated learning	Examples of extended teaching opportunities
Job observation	Operation modes	Job modification
	Human error prevention	Marketing
	Drug manufacturing	Drug discovery
	FDA regulations	Compliance
	Job observation	Communication
	Positive reinforcement planning	

SWOT, strengths, weaknesses, opportunities, threats; FDA, Food and Drug Administration.

1.8.3
Timeline

The timeline for a case study can vary and depends on instructor availability, company needs, instructor goals, and student deliverables. For example, a particular instructor may be available only two weeks in a semester, whereas another instructor may be present the entire semester. If the company is using a case study to "get a job done", for which otherwise the company would have limited or no resources, this is an incentive to tailor the timeline for the case study according to goals and student deliverables rather than to a specific amount of time. The case studies in this book have a length of six to eight weeks, with students meeting for sessions once per week. This study period can be decreased by increasing the number of sessions per week or by altering the assignments. Likewise, the study period can be extended by adding assignments or allowing more time for the execution of the proposed assignments.

1.8.4
Location

Industry immersion learning is optimal when it takes place in industry settings. However, sometimes it is necessary to use less ideal settings, for example, classrooms at universities. The circumstances that will determine the location can be addressed by asking the following questions:

1. Availability:
   (a) Are there a company and an instructor affiliated with the company in the vicinity?
   (b) Is there a conference room or laboratory available at the company site?
   (c) Are there computers or other important technical support for the students?

2. Attainability:
   (a) Is it possible to secure the resources necessary within the defined timeframe?
   (b) Are all legal requirements being met?
3. Practicality:
   (a) Is it possible for the students to arrive at the location in time for class and also to depart from the location in time for other classes?
   (b) Is it possible for students to arrange transportation to the location?
   (c) Is parking available?
4. Optimal environment:
   (a) Is the location conducive to industry learning?
   (b) Is learning dependent on a certain location, i.e., does the learning require a company setting or can it be performed in a university classroom setting?

1.8.5
Teaching Format

Learning about industry includes learning to function in a typical industry atmosphere. Therefore, it is essential that students learn to function well in teams, to listen well, to adopt the necessary flexibility and discipline, as well as to absorb the required technical skills. A variety of teaching formats creates the optimal learning experience. As an example, students need a certain amount of base knowledge before they can perform specific tasks in industry. Such knowledge can be provided in the form of lectures, homework, and team training, or by simple research assignments. More creative assignments often demand a higher level of interaction between industry instructors and team members. This phase is characterized by interactive learning through discussions and participatory problem solving as well as out-of-the-box thinking.

1.8.6
Student Deliverables

When possible, the industry environment should be mimicked as much as possible both in terms of the deliverable and the delivery. Student deliverables such as reports and presentations should resemble those assembled by employees in a company. These can vary widely, and examples include, board presentations, business recommendations, market analyses, business plans, technical documents, management documents, and intellectual property documents. Delivery may be practiced in a formal setting and while wearing business attire to pursue the maximum "learning-by-doing" effect.

1.8.7
Legal Requirements

Industry instructors and their affiliated companies are most often required to protect their intellectual property. This means that students and academic advisors often must sign certain legal documents, for example, confidentiality agreements and agreements relating to transfer of information. Such documents are commonly exchanged between companies. They relate to the information that will be shared, the location and people covered by the agreement, innovation that may take place, and presentation of confidential information. It is important for the establishment of industry immersion education that universities are willing to sign and honor such agreements.

1.9
Publishing of Industry Immersion Case Studies

Industry immersion teaching can initially be a rather resource-consuming endeavor. Being able to re-use teaching materials means conserving resources and thereby potentially increasing the audience. This can be achieved by publishing and disseminating the individual industry immersion case studies. However, industry instructors must ensure that publishing is in compliance with their company's legal regulations and policies.

This book is an example of the willingness and ability of many industry instructors in the Research Triangle Park, North Carolina, to give their time and expertise to students and to find the flexibility and support in their companies to provide such services. Each chapter contains an introduction to a "real-life" topic, the actual case study, followed by an overview of the timeline, teaching plan, and assignments. For most chapters web presentations introducing each topic can be found at www.wiley-vch.de/supplements/. It is my hope that both students and instructors will continue to find this material invaluable for understanding industry topics as well as a constant inspiration to personalizing new educational case studies in their own environments.

2

Integration of Pharmaceutical and Diagnostic Co-Development and Commercialization: Adding Value to Therapeutics by Applying Biomarkers

Michael Stocum

Contents

Industry Immersion Learning. Real-Life Industry Case-Studies in Biotechnology and Business
L. Borbye, M. Stocum, A. Woodall, C. Pearce, E. Sale, W. Barrett, L. Clontz, A. Peterson, J. Shaeffer
Copyright © 2009 WILEY-VCH Verlag GmbH & Co. KGaA, Weinheim
ISBN: 978-3-527-32408-8

2.1
Mission

This case will explore the current state of pharmaceutical and diagnostic development and identify opportunities to optimize product development and commercialization for each industry through co-development of products. Specifically, students will develop a project plan that offers solutions to technical, scientific, and business challenges for the integration of diagnostics in example drug discovery programs.

2.2
Goals

This case will

- provide examples of therapeutic development projects that integrate diagnostics;
- enable students to understand the complexity of developing therapeutics as well as diagnostics;
- offer an opportunity to apply students' existing science and business knowledge to the field of personalized medicine.

2.3
Predicted Learning Outcomes

Students will learn the complexity of developing therapeutics, as well diagnostics. They will be offered the opportunity to apply their learning by designing both technical and business solutions to the issues within co-development of these areas. At the end of the case, students will

- be able to choose a drug target and a molecule (existing small molecule or biopharmaceutical);
- develop a hypothesis-driven targeted drug development program that could result in a personalized medicine product.

2.4
Introduction

In the latter half of the twentieth century and into the beginning of the twenty-first century, tremendous growth occurred in the global pharmaceutical industry (inclusive of biotechnology companies developing and commercializing therapeutics). Accompanying this growth was a dramatic increase in the availability of reliable and generally safe medicines throughout most of the developed world and portions of the developing world. As the mantra of one pharmaceutical company states, these medicines are "helping people do more, feel better and live longer"[1]. Such advancements have resulted in vaccines that have almost eradicated childhood diseases such as polio and measles, replacing extreme fear with an expectation that children will be healthy throughout their early development.

Other infectious diseases viewed as conditions with high mortality rates and rapid progression, such as the acquired immune deficiency syndrome (AIDS), and its causative agent the human immunodeficiency virus (HIV), are now routinely managed to extend infected patients' lives by 10, 15, even 20 years or more. The advancements have also led to medicines that reduce the incidence and spread of diseases such as typhoid fever, leishmaniasis, and malaria in developing countries. In addition, therapies exist today that treat a wide variety of diseases from chronic conditions such as cardiovascular disease and diabetes to acute disease episodes such as sepsis, myocardial infarction, and stroke. Even one of the most challenging human diseases, cancer, is increasingly manageable, resulting in longer life expectancy and improved quality of life.[2]

Similarly, the medical diagnostics industry experienced tremendous growth in the last quarter of the twentieth century and is poised to experience further

1) "Helping people do more, feel better and live longer" is a Service Mark of GlaxoSmithKline.

2) *NCI Cancer Trends Progress Report – 2007.* Available online at: http://progressreport.cancer.gov/index.asp, last accessed 8 December 2008.

growth in the early twenty-first century. While both the pharmaceutical and diagnostics industries are expected to gain from the increased knowledge of the human genome, proteome, and other research within the areas of human biology and disease etiology, the diagnostics industry has the ability to directly "productize" such findings.

An example is the recent finding that the individual level of expression (messenger RNA, or mRNA) from 16 genes predicts whether a woman with breast cancer needs chemotherapy for her cancer or whether she can avoid it while having a relatively high confidence that her chances of survival will be similar or identical to if she had received the chemotherapy.[3] Genomic Health (Redwood City, CA), a company that offers clinical diagnostic laboratory testing services, has commercialized a test, OncoType Dx, based on these findings.

Other examples exist where an improved understanding of the interindividual genetic variation in drug-metabolizing enzymes and vitamin K receptor allows physicians to select the appropriate dosage of the anticoagulant drug S-warfarin.[4] This drug is difficult to dose, and this difficulty results in thousands of patients every year being at risk of adverse events due to the drug.[5]

Approximately 30 years have elapsed since the utility of cholesterol testing was recognized, and today it is assessed during routine medical examinations. Perhaps the biggest utility of cholesterol testing has been to predict risks of cardiovascular disease, particularly atherosclerosis and associated heart attacks and strokes. Medicines have been developed to treat elevated levels of cholesterol, and a corresponding reduction in the incidence of cardiovascular disease has been observed.[6] Additional innovative products have been commercialized by the diagnostics industry enabling tracking of patient's general lipid profile, including triglycerides, high-density lipoprotein cholesterol (HDL), and low-density lipoprotein cholesterol (LDL), resulting in further refinements to the treatment and management of cardiovascular disease.[7,8]

In the post-World-War-II era of healthcare in most of the developed countries, therapeutics and diagnostics have been used in tandem by caregivers to diagnose, treat, and manage disease. People are beneficiaries of this synergy, enjoying improved health and greater productivity throughout increasing life spans.

3) Paik, S., Tang, G., Shak, S., et al. "Gene expression and benefit of chemotherapy in women with node-negative, estrogen receptor positive breast cancer." *Journal of Clinical Oncology* 2006; 24(23):3726–334.

4) Rieder, M. J., Reiner, A. P., Gage, B. F., et al. "Effect of VKORC1 haplotypes on transcriptional regulation and warfarin dose." *New England Journal of Medicine* 2005; 352:2285–2293.

5) Lazarou, J., Pomeranz, B., Corey, P. "Incidence of adverse drug reactions in hospitalized patients – a meta-analysis of prospective studies." *JAMA* 1998; 279(15):1200–1205.

6) "Executive Summary of the Third Report of The National Cholesterol Education Program (NCEP) Expert Panel on Detection, Evaluation, and Treatment of High Blood Cholesterol in Adults (Adult Treatment Panel III)." *JAMA* 2001;285:2486–2497.

7) http://www.labtestsonline.org/understanding/analytes/lipid/glance.html (last accessed 23 September 2008).

8) Ito, M., Cheung, R., Gupta, E., et al. "Key articles, guidelines, and consensus papers relative to the treatment of dyslipidemias – 2005." *Pharmacotherapy* 2006; 26:939–1010.

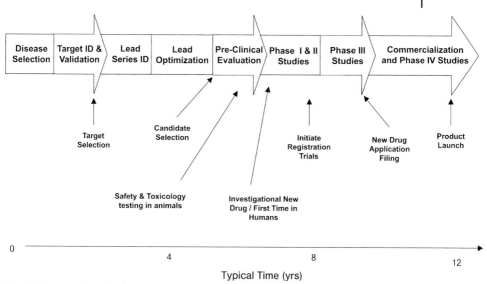

Fig. 2.1 Drug product development cycle.

The advances, however, are coming with increasing cost and broader healthcare, societal, and ethical dilemmas.

2.4.1
Current Environment for Pharmaceutical and Diagnostic Product Development

Pharmaceutical and diagnostic product development consists of many complicated steps; an overview of them is provided in Figure 2.1. Commonly referred to as the "product development cycle", the continuum of drug research, discovery, development, and commercialization begins when an organization selects a disease condition to pursue. The selection of biological targets is a critical first step, as the targets must be relevant in the disease area of interest and must be "addressable", meaning that the target may be altered by a pharmacological agent (a biological or chemical molecule). Once a target is identified, validation is achieved by altering the target and demonstrating through in vitro experiments and some animal models (frequently mice) that the alteration in the target somehow changes the biology of the disease. Following target validation, large libraries of molecules are screened against a validated target to look for molecules that alter the target and the disease in the disease models. A small number or series of molecules demonstrating the best profiles are then optimized to create several molecules that have drug-like properties and address the target. The organization will then select from these one or more candidates to run through safety and toxicology testing in animals and standard laboratory safety tests as required by regulatory agencies. When these tests are completed, an Investigational New Drug (IND) application is filed with the Food and Drug

Administration (in the United States) or similar regulatory body in other countries or regions of the world.

Following this submission, and in the absence of regulatory holds on further research, the organization is free to pursue the first studies in humans. The molecule is studied at low doses as established from safety and toxicology studies in animals, then given repeatedly at larger doses so that clinical scientists may assess the properties of the molecule in humans. Assuming the drug is found to be safe and has the appropriate pharmacological properties, including pharmacokinetics (the activity or fate of the molecule in a living body) and pharmacodynamics (how the molecule interacts with the disease target), the organization will advance the molecule into additional clinical trials and test its effectiveness (efficacy). Generally, phase II studies are the point at which companies may begin evaluating the efficacy of the molecule, and, assuming positive results, will continue by conducting larger studies for registration. These phase III studies are designed to test the safety and efficacy of the molecule in large patient groups (several hundreds to thousands of patients).

When an organization feels it has developed statistically significant data to support product claims in regard to the molecule, it will file a New Drug Application (NDA) (if in the United States, or a similar document in other regions). Regulatory agencies will consider this application and whether the clinical trial data support the claims sought by the organization. If it does, the drug label is approved, and the product quickly begins its commercial phase; the drug is sold and patients are treated outside clinical trials. If it does not, new studies may be necessary to achieve approval. Once the drug is approved and commercial distribution begins, additional studies may be conducted to expand the label claims, develop new indications in other areas of disease, or simply provide marketing organizations with additional and more compelling data to increase sales. These are frequently referred to as phase IV trials.

Like the drug product development cycle, the diagnostic product development cycle can be lengthy when one considers the starting point as discovery of a biomarker (Figure 2.2). However, most diagnostic products are developed and distributed by companies that typically only engage in diagnostic product development after biomarkers have already been found that demonstrate significant clinical validity. For instance, the biomarkers are present and change appropriately in the context of changing disease. At this stage, multiple companies may pursue product development on their proprietary platforms. At the same time, Central Reference Laboratories often begin offering service testing for the biomarker (or biomarkers) long before a product is commercialized by a traditional diagnostics company. These traditional diagnostics companies refine the use of the biomarkers on their platform and conduct clinical trials to register the product (generally in the United States; other countries vary in their regulatory requirements). Upon successful completion of the registration studies and subsequent regulatory approval or clearance, the product may be launched and sold as an *in vitro* diagnostic kit. The entire cycle is generally an 8- to 10-year program, but the companies active in the diagnostic product development space

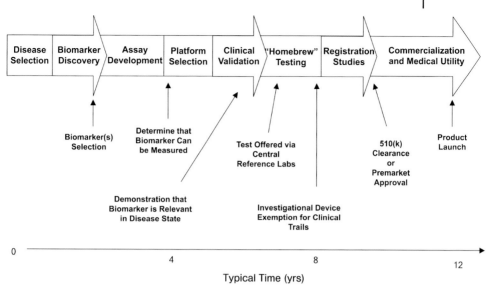

Fig. 2.2 Diagnostics product development cycle.

typically complete the product cycle in 4–6 years, depending on which bio-markers are chosen and the relative characteristics of that disease area.

For pharmaceutical companies, challenges exist in several forms, including:

- Upcoming patent expirations (see Chapter 5) allowing generic competition on high-profit branded products with large revenues (> $1 billion)
- Increasingly long product development cycles (see Figure 2.1)
- Escalating costs for research and development
- Rising failure rates (drug candidates that never achieve product label approval), often later in the development cycle with greater cost
- Heightened regulatory oversight, especially with regard to safety
- Continuing threat to restrain patentable dicoveries; risk of limiting current and future intellectual property rights that protect market exclusivity
- Uncertainty regarding drug reimbursement by payers, especially on products with only marginal improvements in drug efficacy
- General public mistrust, mistrust, suspicion, and skepticism regarding pharmaceutical industry marketing practices and product claims

Of these challenges, the one over which pharmaceutical companies have the greatest control is the product development cycle. While overall costs might rise due to ever-rising labor costs and the expense of acquiring novel technology to support development efforts, major costs for the industry, on average, are due to late-stage product failures. Such failures often occur during phase III, which is generally the final stage of testing prior to regulatory approval for marketing product label claims. By this stage companies have typically invested hundreds of millions of dollars in disease research, drug discovery, and development of the drug.

Failure costs are amortized across the portfolio and balanced by successful drug programs, i.e., those achieving successful product registration. The amortized cost of developing a new drug was estimated at $802 million in the early 2000s.[9] More recently, the cost of developing a new biotechnological product has been estimated at $1.2 billion.[10] In stark contrast, a successful targeted therapeutic development program in a carefully defined disease patient population that achieves clinical endpoints early in the development process may cost less than $100 million to reach successful product registration.[11] Understanding the aforementioned dynamics, pharmaceutical companies are increasingly refining the use of their development resources to achieve success in a higher percentage of programs and curtailing those development programs that are unlikely to succeed at an earlier point in the development cycle.

Similar to the pharmaceutical industry, the diagnostics industry has endured its own challenges. These include:

- Comparatively low profit margins on products sales (i.e., compared to pharmaceutical products)
- A lack of innovative, proprietary, novel products
- Reimbursement levels based on earlier-generation cost of product versus actual value of product reimbursement rates versus value-based reimbursement
- Escalating costs for development and validation of novel products
- Scarcity of appropriately collected and preserved samples with annotated clinical information (samples must be collected under patient-informed consent and preserved/banked)
- Intellectual property concerns, including patent protection that may be circumvented for products and need for multiple licenses or "patent stacking" requirements for freedom to operate and develop products (in "stacking," multiple patents are needed for different aspects of a single innovation, thus forcing several royalty applications and payments)
- Uncertainty about new regulation of laboratory service tests, for example, the In Vitro Diagnostic Multivariate Index Assays draft guidance (see http://www.fda.gov/cdrh/oivd/guidance/1610.pdf)

Unlike the pharmaceutical industry, which historically has enjoyed high margins on its products as well as overall larger revenues, the diagnostics industry often lacks the financial clout and resources to drive change in its marketplace. As a result, the diagnostics industry is largely dominated by companies that sell

9) DiMasi, J., Hansen, R. W., Grabowski, H. G. "The price of innovation: new estimates of drug development costs." *Journal of Health Economics* 2003; 22:151–185.

10) Tufts Center for the Study of Drug Development press release, http://csdd.tufts.edu/ NewsEvents/NewsArticle.asp?newsid=69 (last accessed 23 September 2008).

11) Personal communication from former Novartis employee wishing to remain anonymous regarding the clinical development costs of imatinib (Gleevec) from first time in humans to new drug approval by FDA.

high volumes of (minimally profitable) products which generate test results when used on that company's sophisticated laboratory analyzers and associated equipment.

The characteristics of each of these industries, combined with the challenges each faces and their important contribution to the successful diagnosis, treatment, and monitoring of disease by the medical profession, offer unique synergies. One of these is the co-development of products between the industries to their mutual benefit. This subject has received significant investment from both industries and is the subject of this case study, with a focus on how the pharmaceutical industry is utilizing this approach. Co-development often results in products that have a better defined clinical benefit, are a clearer value proposition to the paying community, and result in improved patient outcomes. These products and related means of assessing and treating patients are collectively referred to as "personalized medicine".

2.4.2
Potential Solutions to the Challenges Confronting Pharma

The twenty-first century was ushered in by one of the greatest biological research accomplishments in human history – the sequencing of the human genome. Approximately three billion base pairs were sequenced and 20 000–25 000 genes elucidated. For over a decade, research-based pharmaceutical companies have sought to capitalize on the sequencing of the human genome. The efforts have focused especially on basic disease research, identification of novel putative drug targets, and expanding the depth of knowledge on existing drug targets. Concurrently, academic and government-sponsored basic and applied clinical research efforts have steadily identified novel biomarkers that have potential as future diagnostic products. Such products may be useful in combination with drugs active against new and existing drug targets. This offers the possibility of a more detailed molecular definition of disease. It also enables the identifying and targeting of specific patient populations with drugs that have an increased likelihood of response.

2.4.2.1 Genomics and Proteomics, Metabolomics, and "Other -omics"

Throughout the so-called "genome era", companies have utilized a variety of technologies for assessing the genome from DNA and RNA to the protein products encoded by the genes. Additionally, technological advances have provided deeper insight into metabolite and cellular profiles at the molecular level, enabling a broad picture of the underlying biology, often referred to as "systems biology." The technologies that contribute information to a systems biology approach are collectively referred to as "-omics technologies". The -omics include genomics, proteomics, metabolomics, glycomics (study of carbohydrates), lipomics (study of lipids), and cellomics. Table 2.1 outlines several example technologies and their application to systems biology research.

Table 2.1 Application of "-omics" technologies to systems biology research.

Type of -omic	Technology	Purpose
Genomic	Comparative genomic hybridization	Identification of chromosomal aberrations including gene duplication, insertions, or deletions
Genomic	DNA sequencing	Reading of individual base pairs in the sequence of DNA molecules
Genomic	DNA methylation testing	Identification of regions of methylated DNA that can serve to promote or attenuate gene expression
Genomic	mRNA expression analysis	Evaluation of mRNA levels provides insights into possible biological pathway activation
Proteomic	Protein analysis	Assessment of protein levels often directly relates to cellular activity
Metabolomic	Metabolite profiling	Assessment of changes in metabolites to provide indication of changes in body function
Lipomics	Lipid profiling	A subset of metabolomics, lipid profiles are useful in understanding a wide range of human physiological conditions
Glycomics	Carbohydrate profiling	Also a subset of metabolomics, carbohydrate profiles are being evaluated to better understand cellular metabolism
Cellomics	Cell signaling assessments	Often evaluated to understand cellular function and communication

2.4.2.2 Translational Research

The application of what is learned in basic research to clinical (applied) research is generally referred to as translational research. Indeed, the term "translational" is frequently applied to many stages of research and product development and has its roots in the belief that applying knowledge that transcends specific disciplines can better enable and inform each individual discipline. Relative to pharmaceutical development, translational research is specifically dedicated to capitalizing on knowledge acquired during the basic research phase and applying that knowledge to the clinical development stage that includes pharmaceutical drug trials. Increasingly, translational research is utilizing various -omic technologies to better understand drug dosing, patient response, and disease prognosis (or predicting the disease course for the patient). Translational research also has a powerful effect on basic research in that the knowledge gleaned in the clinic can be used to improve basic research planning and execution.

2.4.2.3 Biomarkers

A biomarker is defined as "a characteristic that is objectively measured and evaluated as an indicator of normal biological processes, pathogenic processes, or pharmacologic responses to a therapeutic intervention".[12] The measurement of biomarkers in translational research and clinical development has significant implications for pharmaceutical and diagnostic product development. Results from such studies frequently provide clinical validation data that support investment in large product registration studies for both drugs and diagnostics. Biomarker research has a wide range of applications that support drug discovery, development, and commercialization, including the following:

- Improved understanding of disease etiology and biology
- Early identification of potential drug failures prior to significant investments
- Accelerated product development [see section on imatinib (Gleevec), this chapter]
- Identification of patient populations highly likely to respond to the drug [see section on trastuzumab (Herceptin), this chapter]
- Monitoring of disease progression and treatment response
- Expansion of commercial markets through early detection of disease at a presymptomatic stage or identification of additional populations that will benefit from treatment

Biomarkers are not new to biomedical research and they have frequently been used to support the development of drugs in several markets. These include serum cholesterol levels (used as a biomarker in the development of the statin class of drugs for cardiovascular disease), blood glucose levels of hemoglobin A_{1c} (used in the development of several drug classes for type 2 diabetes, which are driven by the measurement of this biomarker), and HIV-1 RNA levels and CD4+ cell counts, used as surrogate endpoints in the development of multiple drugs for the treatment of HIV-1 (CD4+ cells are a marker of immune system competency). But perhaps no therapy area stands to benefit from biomarker-supported drug development as much as oncology. The broad-reaching efforts of research into the biology of cancer combined with the historically low response rates to drugs used to treat this disease have led to both a need for improved drug development and the knowledge to make a significant impact.

2.4.3
Drug Development for Targeted Cancer Therapies

2.4.3.1 Tamoxifen in Estrogen-Receptor-Positive Breast Cancer
An example of targeted therapies in cancer is the development of estrogen receptor (ER) antagonists such as tamoxifen.[13] Estrogen, a hormone normally produced by the ovaries of premenopausal women, can drive the growth of breast tumors when cells overexpress estrogen receptor. Such tumors are re-

12) Biomarkers Definitions Working Group. "Biomarkers and surrogate endpoints: preferred definitions and conceptual framework." *Clinical Pharmacology & Therapeutics* 2001; 69(3):89–95.

13) Hortobagyi, G. "The status of breast cancer management: challenges and opportunities." *Breast Cancer Research and Treatment* 2002; 75(Suppl 1):S61–65.

ferred to as ER-positive breast cancer and are generally treated with drugs such as tamoxifen. The ability to test by immunohistochemistry for ER status in breast cancer tissue enables the identification of patients most likely to benefit from this type of cancer therapy. For those who are ER-negative and unlikely to benefit from tamoxifen, other therapeutic options may be immediately pursued, such as radiation, chemotherapy (toxic compounds used against aggressive and/or metastatic cancers), or biological therapies.

2.4.3.2 Trastuzumab in Breast Cancers Overexpressing Her2

Another example of a targeted therapeutic for breast cancer is the biological drug trastuzumab (Herceptin; Genentech, San Francisco, California, USA), a recombinant DNA-derived humanized monoclonal antibody against the Her2/neu protein (Her2). The Her2 protein is overexpressed on the cell surfaces of approximately 20%–25% of all breast cancers.[14] The overexpression of protein may be directly identified through immunohistochemical (IHC) testing of breast cancer tissue for Her2 by the HercepTest or other similar diagnostic products. Her2 overexpression may also be identified by fluorescence in situ hybridization (FISH), which involves the use of nucleic acid probes that bind directly to the gene on the chromosome that is responsible for production of the Her2 protein. When multiple copies of this gene are present repeatedly on individual chromosomes, the result is overexpression of Her2 protein on the surface of tumor cells. Tumors overexpressing Her2 have up to a 40% response rate to trastuzumab. Currently, researchers are working to better understand why the other 60% still do not respond despite overexpression.

The clinical trials of trastuzumab were originally conducted in previously treated breast cancer patients whose tumors overexpressed Her2, as measured by IHC staining of formalin-fixed paraffin-embedded tissue biopsy material from the patient. Although response rates to trastuzumab vary by disease stage and other unknown factors, it is clear from the clinical development and subsequent studies that patients who do not overexpress Her2 generally exhibit no response to trastuzumab. These observations led the US Food and Drug Administration (FDA) to require the development of an *in vitro* diagnostic IHC test for Her2 expression for use in identifying patients who overexpress Her2 and are likely to benefit from trastuzumab therapy.

Herceptin received regulatory approval on 25 September 1998 from the FDA for use as first-line therapy in combination with paclitaxel (one of several chemotherapy agents frequently used in the treatment of breast cancer) and as a single agent in second- and third-line therapy for patients with metastatic breast cancer who have tumors that overexpress the HER2 protein. Concurrently, Dako (Copenhagen, Denmark) received premarket approval (PMA) from the FDA for an IHC test that identifies Her2 overexpression and, hence, fulfills the role of

---

14) Menard, S., Pupa, S. M., Campiglio, M., et al. "Biologic and therapeutic role of HER2 in cancer." *Oncogene* 2003; 22(42):6570–6578.

the initial test for selection of metastatic breast cancer patients for trastuzumab therapy. This was the first simultaneous approval of a therapeutic and a diagnostic as companion products. This product combination and other diagnostics subsequently approved (two DNA-based tests by FISH and one other for Her2 overexpression by IHC staining) are frequently touted as examples of personalized medicine.

2.4.3.3 Imatinib (Gleevec) in Chronic Myelogenous Leukemia and Gastrointestinal Stromal Tumors

A third example of a targeted cancer therapy is the tyrosine kinase inhibitor imatinib mesylate (Gleevec or Glivec; Novartis, Basel, Switzerland). Imatinib works by specifically inhibiting enzymatic activity in several kinase domains of proteins, including BCR-ABL, the abnormal tyrosine kinase created by the Philadelphia chromosome abnormality in chronic myeloid leukemia (CML). Imatinib also inhibits other kinases, including receptor tyrosine kinases for platelet-derived growth factor (PDGF) and stem cell factor (c-kit) as well as other cellular events mediated by stem cell factor. Specifically, imatinib has been shown to inhibit proliferation and induction of apoptosis (programmed cell death) in gastrointestinal stromal tumors (GIST) whose cells express an activating *c-kit* gene mutation.

Diagnostic tests played a key role in the development of imatinib for each cancer type. For example, CML is diagnosed through identification of the genetic translocation that generates the BCR-ABL fusion protein that is inhibited by imatinib. Generally identified by FISH, this protein results from a translocation event, designated t(9;22), when the genetic material of chromosome 9 is exchanged with chromosome 22. It results in one chromosome 9 longer than normal and one chromosome 22 shorter than normal, which is referred to as the Philadelphia chromosome. The *BCR-ABL* gene is located on the Philadelphia chromosome. When the Philadelphia chromosome is absent, the patient is considered cured of CML. Although infrequent, some cures have been reported. More common is a prolonged efficacious response, followed by gradual resistance to imatinib. Resistance may be measured by DNA sequencing of the ABL kinase domain and comparing the results to known wild-type and mutant ABL kinase sequences.[15] Understanding the resistance profile provides insights into possible therapeutic combinations with other novel BCR-ABL kinase inhibitors, such as dasatanib (Sprycel; Bristol-Myers Squibb, New York, New York, USA).

Diagnostic tests also play a role for the potential use of imatinib in GIST, where activating mutations in the kinase domain of c-kit predict for response to the agent. However, the diagnostic is far less critical in GIST than in the Her2

15) Weisberg, E., Manley, P. W., Cowan-Jacob, S. W., Hochhaus, A., and Griffin, J. D. "Second generation inhibitors of BCR-ABL for the treatment of imatinibresistant chronic myeloid leukaemia". *Nature Reviews Cancer* 2007; 7:345–356. Available at: http://www.cmlsupport.org.uk/?q=system/files/Mechanisms+of+Resistance+-+Imatinib.pdf (last accessed 23 September 2008).

example, where the majority of breast cancer patients do not overexpress Her2. Most GIST do express c-kit and have activating mutations present, which means that the use of a test in combination with the therapeutic only slightly improves the response rates. As in CML, resistance mutations have been identified in GIST following imatinib therapy. Patients who develop such resistance have gone on to show response to another novel c-kit kinase inhibitor, sunitinib (Sutent; Pfizer, New York, New York, USA).

2.4.3.4 Other Targeted Therapies

Additional targeted agents are currently being marketed, including both biological and small-molecule drugs. Two key biologicals recently approved for marketing in the United States are bevacizumab (Avastin; Genentech), a vascular endothelial growth factor (VEGF) inhibitor, and cetuximab (Erbitux; Imclone Systems, New York, New York, USA), an epidermal growth factor receptor (EGFR) inhibitor. Cetuximab was developed in patients whose tumors overexpress EGFR as measured by IHC staining. However, subsequent studies have not demonstrated a correlation between protein overexpression and response to cetuximab. Biomarkers have yet to be identified that predict response to bevacizumab.

Two small molecules recently approved for marketing in the United States are erlotinib (Tarceva; OSI Pharmaceuticals, Melville, New York, USA/Genentech/Roche, Basel, Switzerland), an EGFR inhibitor, and sorafenib (Nexavar; Onyx Pharmaceuticals, Emeryville, California/Bayer Pharmaceuticals, now Bayer-Schering, West Haven, Connecticut, USA), which inhibits a wide range of kinases. Erlotinib was developed in non-small-cell lung cancer (NSCLC) and has been evaluated in other tumor types. Sorafenib was developed in renal cell carcinoma and is under evaluation in other tumor types. Efficacy rates for both drugs are low (as compared to imatinib in CML and GIST), and neither has a robust biomarker profile. However, both have demonstrated a survival benefit for patients with the targeted tumor type versus the current standard of care, and as a result they have received approval.

2.4.4
Specific Example of Lapatinib (Tykerb)

2.4.4.1 Leveraging Biomarkers and Diagnostics to Accelerate Drug Development

As mentioned previously in this chapter, biomarkers and diagnostic tests may be leveraged to accelerate drug development. Lapatinib (Tykerb; GlaxoSmith Kline, London, UK), a drug that targets both Her2 and EGFR, received FDA approval in March 2007. Lapatinib is indicated for women with breast cancer who have received prior therapy with other cancer drugs, including an anthracycline, a taxane, and trastuzumab (Herceptin).[16] The drug is also under evaluation in

---

16) "FDA Approves Tykerb for Advanced Breast Cancer Patients." FDA press Release, March 2007. Available at: http://www.fda.gov/bbs/ topics/NEWS/2007/NEW01586.html (last accessed 23 September 2008).

other stages of breast cancer and other tumor types, having shown promise in a variety of cell lines (preclinical testing) as well as anecdotal reports of responses from patients with non-breast-cancer tumors during phase I studies.

To understand how biomarkers have impacted on the development of lapatinib, one needs to review the history of the preclinical program and early clinical studies. Lapatinib was discovered in the late 1990s by cancer researchers at GlaxoWellcome, a predecessor to the company that currently markets the drug, GlaxoSmithKline (GSK). By early 2000, lapatinib was just undergoing the final rounds of preclinical and animal toxicity testing. During this period, clinical scientists in academia and industry were developing a translational medicine program that would provide the foundation for an accelerated drug development program. This included evaluating the drug in human cell lines for its effect on biomarkers in the epidermal growth factor (EGF) family of receptor tyrosine kinases (ErbB) pathway, specifically Her2 and EGFR signaling and apoptotic pathways.

Scientists were also poised to take advantage of novel biomarker imaging and quantification technologies being applied to traditional IHC testing. Such technologies were validated in 2000 using tumor xenograft samples [17] and accelerated the application of these assays for use on human clinical samples.

Lapatinib was first tested in humans in January 2001. Over a year and a half, three key phase I lapatinib studies were performed to evaluate safety and tolerability as well as pharmacokinetic properties across several drug dosage levels. While these parameters are very typical objectives of understanding in early pharmaceutical trials, lapatinib clinical researchers had the vision to extract additional information. A range of biomarker testing was incorporated into clinical protocols. This allowed collection and testing of human clinical samples before and after treatment with lapatinib, using several of the same assays that were validated on the tumor xenograft samples. Biomarker data from these studies were then compared against observed clinical results. These same data were also used to demonstrate specific changes in selected biomarkers, for instance, a 75% reduction of phosphorylated Her2 protein levels (so-called activated Her2, because it is actively signaling). When statistical models were applied to the collected clinical and biomarker data, several of the biomarkers were found to predict clinical outcomes in cancer patients treated with lapatinib. Examples are a marker of functional apoptosis, TUNEL; two cell surface receptors, ErbB2 and insulin-like growth factor receptor 1 (IGF-1R); a known ligand for these receptors, transforming growth factor-a (TGFa); and two downstream signaling pathways proteins, phosphorylated S6 kinase (pS6K) and phosphorylated Erk1/2.

The demonstrated reduction of signaling in the ErbB pathways (especially Her2 and EGFR) targeted by lapatinib was an objective of the studies. Conclusively, the project team was able to convince senior management to invest in evaluation of this drug in a variety of human tumor types in phase II studies.

---

17) A xenograft in this context is a portion of a human tumor grafted onto a nude mouse; without a functional immune system, the mouse does not reject the human tissue.

There was such confidence in the success of this development program, largely driven by biomarker data, that a decision was made to accelerate investment in phase III registration studies in breast cancer simultaneously with the phase II programs. This is highly unusual because companies frequently await the results of phase II studies before initiating costly phase III studies. However, in this case, the phase III studies, initiated in early 2003, provided the data that supported the NDA submitted by GSK in 2006. As previously mentioned, FDA approved marketing of lapatinib, branded as Tykerb, in March of 2007, indicated for breast cancer patients who have received prior therapy. Clinical researchers continue to pursue development in early stages of breast cancer as well as in other cancers, especially those with a subpopulation of tumor types driven by the ErbB signaling pathways.

Under the current label indication, identification of cancer patients most likely to respond to lapatinib therapy may be based on anatomical tumor location (i.e., breast cancer), prior treatment history, and Her2 overexpression. Two of these three are generalized patient selection criteria that might apply to a variety of drugs, while Her2 is a more specific biomarker typically found in aggressive breast cancers. While these current selection criteria are sufficient for modest responses to lapatinib, they are unlikely to support broader use of lapatinib, especially in tumors other than breast cancer. Additionally, they do not enable higher efficacy rates in breast cancer, particularly in earlier stages of breast cancer, where a generally well-tolerated drug such as lapatinib could undergo widespread and long-term usage. Such usage frequently garners in excess of $1 billion in annual sales and creates a drug with blockbuster status.

2.4.4.2 Potential to Enhance Commercial Success with Companion Diagnostics

Although the initial lapatinib biomarker data suggested that some of the biomarkers had potential utility for improving patient selection or predicting response to therapy, the data were not strong enough to induce commercial entities to develop these biomarkers into diagnostic tests to support lapatinib. Now with lapatinib on the market, there are many ways in which tests may be used to improve and expand the utilization of this therapeutic. The three key approaches with potential for enhancing commercial success of lapatinib through the use of companion diagnostics are:

- Improving patient selection criteria in earlier stages of breast cancer
- Monitoring patient responsiveness to lapatinib, allowing optimized dosing and combination therapy
- Identifying patients with non-breast-cancer tumors who could benefit from lapatinib

Clinical researchers have continued their efforts to understand at the molecular level how tumors respond to lapatinib. The biomarker patterns are often termed "molecular signatures". Simultaneously, ongoing research efforts in academia and industry to improve the understanding of the molecular signature of tumors are beginning to show promise. As the two types of data converge, opportunities to develop companion diagnostics are emerging. Companies choosing to develop such tests may utilize their own resources or seek a development partnership with the company that developed and markets the drug (GSK in the case of Tykerb). Developing these companion diagnostic products is a core component to the "personalization" of medicine.

2.4.5
Personalized Medicine

"Personalized medicine" is often defined as *the integration and application of an individual's unique healthcare information to predict, prevent, diagnose, and treat disease, as differentiated from traditional medical practice supported by population-based information.* However, the term may be understood differently depending upon the audience. For example, physicians may suggest that personalized medicine has always been in practice for their patients, regardless of how the information to treat and manage the patients is acquired. The distinction today is that more molecular information is being utilized to treat and manage patients than ever before. Until recent advances in genomics and molecular medicine were made, physicians relied solely on clinical observations, population-based observations, and established clinical chemistry markers measured in patient samples. Now, physicians increasingly are utilizing genomic information and novel molecular diagnostic tests in the treatment and management of their patients. This practice is frequently referred to as "personalized medicine" and is being implemented in selected diseases, including cancer. Many of the targeted therapies for cancer are amenable to a personalized medicine treatment approach.

As novel therapeutics are developed in targeted populations, identified by a molecular signature, data will be available, in tandem with the therapeutic development program, to treat patients with a personalized medicine approach. From the disease perspective, as researchers uncover specific molecular signatures of diseases they will be able to utilize appropriate therapeutics with matching, or at least near-matching, complimentary diagnostic tests and profiles for these molecular signatures. This approach to treating patients holds the promise to optimize the use of therapies, create greater healthcare value, and, most importantly, improve patient care. [18]

2.5
Case Scenario

In this case, students should receive information relating to the development of biomarkers and therapeutics and the value of an integrated approach to drug and diagnostic product development ("companion products"). Additionally, an

18) http://www.personalizedmedicinecoalition.org/
communications/TheCaseforPersonalizedMedicine_
11_13.pdf (last accessed 23 September 2008).

overview of the rationale for how such research efforts lead to the development of novel diagnostic and therapeutic agents and the intrinsic link between the two product types should be given. An example program should be shared with the students, including the supporting science and business rationale. Students are expected to develop a biomarker-driven development program and accompanying personalized medicine plan, including financial projections, where applicable.

2.6
Timeline

Session 1: Introduction of Therapeutic and Diagnostic Product Development
　　　　Case Outline and Objectives
　　　　Development of Technical and Scientific Hypotheses
Session 2: Overview of Challenges and Potential Solutions to the Challenges for the Pharmaceutical and Diagnostics Industries
Session 3: Overview of Drug and Diagnostics Development for Targeted Cancer Therapies
　　　　Overview of a Generic Product Profile
　　　　Development of Product Profile
　　　　Personalized Medicine
Session 4: Development of Integrated Therapeutic and Diagnostic Co-Development Plan
Session 5: Student Work
Session 6: Student Presentations

2.7
Study Plan and Assignments

2.7.1
Session 1

2.7.1.1 Presentations

Drug and Diagnostic Product Development (Presentation #1)
Presentation of case objectives and outline of the final case report structure:

- Introduction
- Objectives
- Scientific rationale for developing integrated tests and medicine for the drug and diagnostic co-development area
- Current examples
- Project plan and timelines

- Business rationale for developing integrated tests and medicine [Strengths, Weaknesses, Opportunities, and Threats (SWOT) analysis)] for the drug and diagnostic co-development area
- Business and alliance management issues
- Impact on therapeutic – model of drug usage
- Recommendations
- Conclusions

2.7.1.2 Assignment #1
Read:

Introduction to this chapter.

Lesko L. J., Atkinson A. J. "Use of biomarkers and surrogate endpoints in drug development and regulatory decision making: criteria, validation, strategies." *Annual Review of Pharmacology and Toxicology* 2001; 41:347–366.

Form groups. Each group should select a target area for biomarker development, complete a technical evaluation, and develop scientific hypotheses regarding which markers might be most useful.

2.7.2
Session 2

2.7.2.1 Presentations
Student presentations of developed hypotheses and discussion of the technical challenges presented.

Challenges in Drug Discovery (Presentation #2)

Biomarkers (Presentation #3)

2.7.2.2 Assignment #2
Develop solutions for addressing the presented technical challenges.

Read:

Meade P. D., Keeling P. "How to make Rx-Dx alliances work." *Pharmaceutical Executive* July 2003; J48–60.

Frank R., Hargreaves R. "Clinical biomarkers in drug discovery and development." *Nature Reviews Drug Discovery* 2003; 2:566–580.

Complete scientific hypotheses.

2.7.3
Session 3

2.7.3.1 Presentations
Developing Personalized Medicine Products (Presentation #4)

Discussion of targeted cancer therapies (especially Tykerb)

Student presentations of solutions to technical challenges and presentations of final scientific hypotheses.
Test Product Profile (Presentation #5)

2.7.3.2 Assignment #3
Read:

Dancey, J. E., Freidlin, B. "Targeting epidermal growth factor receptor – are we missing the mark?" *Lancet* 2003; 362:62–64.

Little S., Blair E. D. "Pharmacodiagnostics: Technologies, Competition, and Market Models." Cambridge Healthtech Associates, *Life Science Reports*, 2005

Molina, M. A., Baselga, J., Clinton, G. M., et al. "NH_2-terminal truncated HER-2 protein but not full-length receptor is associated with nodal metastasis in human breast cancer." *Clinical Cancer Research* 2002; 8: 347–353.

Xia, W., Liu, L., Ho, P., Spector, N. L. "Truncated ErbB2 receptor (p95^{ErbB2}) is regulated by heregulin through heterodimer formation with ErbB3 yet remains sensitive to the dual EGFR/ErbB2 kinase inhibitor GW572016." *Oncogene* 2004; 23:646–653.

Develop a test product profile that will enable commercial assessment of the hypotheses developed in Session 2 and offer enhanced diagnostic and therapeutic co-development opportunities.

Test evaluations can be set out as shown in Figure 2.3.

Review the project that is integrating diagnostic tests in the development of a therapeutic, in light of a test product profile. If no data are available, identify suitable characteristics and define parameters which seem both optimal and realistic.

2.7.4
Session 4

2.7.4.1 Presentations
Student presentations of product profiles and discussion of how to modify these.

2.7.4.2 Assignment #4
Complete modification of product profiles.

Develop a fully integrated plan for diagnostic and therapeutic co-development, including technical and commercial milestones, cost projections, and other details as discussed.

Table 2.2 Twenty key factors to be specified in a test product profile.

1. Indications for use (what patient population should receive the test and what does the information mean for the testing population?)
2. Clinical utility (how is use of the test expected to impact medical practice?)
3. Clinical gold standard (how are patients currently managed and what is the prevalence of the condition being measured?)
4. Correlation with gold standard (what should the treatment population look like after implementation of the test?. What will be the new prevalence of the condition/therapeutic event, etc.?)
5. Positive predictive value (PPV) (defined as all those who test positive and have the condition (true positive) divided by all those who test positive)
6. PPV rationale (PPV is the ability of the test result, in a comparison with the clinical gold standard, to correctly include a patient in the group with the condition for which the test is indicated)
7. Negative predictive value (NPV) (defined as all those who test negative and do not have the condition (true negative) divided by all those who test negative)
8. NPV rationale (NPV is the ability of the test result, in a comparison with the clinical gold standard, to correctly exclude a patient from the group with the condition for which the test is indicated)
9. Marker analyte and assay type (target analyte type: protein, nucleic acid, clinical chemistry marker, etc.)
10. Sample type (whole blood, plasma, serum, urine, tissue biopsy, cerebrospinal fluid, synovial fluid, lavages, etc.)
11. Sampling frequency (when are samples to be taken and how often are samples collected?)
12. Time to clinical delivery of result (how long is it from sample acquisition to delivery of clinical results?)
13. Test location (where is the testing conducted: near patient, physician's office lab, central reference laboratory, hospital laboratories, etc.?)
14. Assay validity (accuracy, sensitivity, and specificity – and impact on PPV/NPV) (see Appendix)
15. Cost of goods sold (COGS) (how much will the test cost to create?)
16. Overall cost and pricing (in addition to COGS, how much additional cost is incurred in bringing the product to market and promoting it? What will be the pricing strategy?)
17. Reimbursement (who will pay for the test, how much, and what are the mechanisms for reimbursement?)
18. Markets (in which geographical locations will the test be distributed?)
19. Intellectual property (IP) (which patents are key to freedom to operate and commercialize? What patents need to be defended to maximize market potential? Are there any additional filings that will improve the IP position?)
20. Timelines

Disease		Test result		
		positive	negative	total
	present	a	c	a+c
	absent	b	d	b+d
	total	a+b	c+d	a+b+c+d

Sensitivity = $\dfrac{\text{All testing positive}}{\text{All those with the condition}}$ $= \dfrac{a}{a+c}$

Sensitivity is a measure of how good the test is at detecting those who have the condition. In other words how well a test detects a true positive result. It is the probability that a patient with the condition has a positive test result.

Specificity

Specificity = $\dfrac{\text{All testing negative}}{\text{All those without the condition}}$ $= \dfrac{d}{b+d}$

Specificity is a measure of how good the test is in detecting those individuals who do not have the condition. In other words, how well a test detects a true negative result or the probability that a patient without the condition has a negative test result.

Disease		Test result		
		positive	negative	total
	present	True positive	False negative	Diseased
	absent	False positive	True negative	Healthy
	total	All Test positives	All test negatives	n

Positive predictive value

Pos PV   = $\dfrac{\text{All testing positive \& with the condition (true positive)}}{\text{All testing positive}}$

= $\dfrac{a}{a+b}$

Predictive values measure how valuable a test is in practice. The Positive PV of a test is the probability of actually having the condition given that the test result is positive.

Negative predictive value

Neg PV   = $\dfrac{\text{All testing negative \& without the condition (True negative)}}{\text{All testing negative}}$

= $\dfrac{d}{c+d}$

The Negative PV of a test is the probability of not having the condition given that the test result is negative.

Fig. 2.3 Test Performance Definitions

2.7.5
Session 5

2.7.5.1 **Assignment #5**
Continue work on the co-development plan.
   Develop an oral presentation (slide show) to address the scientific and techni-
cal requirements and the commercial issues.

2.7.6
Session 6

2.7.6.1 **Presentations**
Student presentations of a fully integrated co-development plan.

Acknowledgment

I wish to acknowledge Lisbeth Borbye for her encouragement, and support of
my writing as well as her dedication to the educational program at North Carolina State University that created the opportunity to write this chapter; Neil Spector, MD, and Paul Meade, MSc, for their contributions in teaching the original
case that formed the basis for this chapter. I also wish to acknowledge Paul
Meade and Edward Blair, PhD, for their comprehensive review of this case
manuscript and their insightful commentary. Finally, I thank my current and
former professional colleagues for helping me develop the broad understanding
of this field, and, lastly, I thank my family for their love and support, without
which none of this would be possible.

Resources

Dancey, J. E., Freidlin, B. "Targeting epidermal growth factor receptor – are we missing the mark?" *Lancet* 2003; 362:62–64.

Frank, R., Hargreaves, R. "Clinical biomarkers in drug discovery and development."
Nature Reviews Drug Discovery 2003; 2:566–
580.

Lesko, L. J., Atkinson, A. J. "Use of biomarkers and surrogate endpoints in drug development and regulatory decision making: criteria, validation, strategies." *Annual
Review of Pharmacology and Toxicology*
2001; 41:347–366.

Little S., Blair E. D. "Pharmacodiagnostics:
Technologies, Competition, and Market
Models." Cambridge Healthtech Associ-
ates, *Life Science Reports*, 2005

Meade, P. D., Keeling, P. "How to make Rx-
Dx alliances work." *Pharmaceutical Executive* July 2003; 48–60.

Molina, M. A., Baselga, J., Clinton, G. M., et
al. "NH$_2$-terminal truncated HER-2 protein but not full-length receptor is associated with nodal metastasis in human
breast cancer." *Clinical Cancer Research*
2002; 8:347–353.

Xia, W., Liu, L., Ho, P., Spector, N. L. "Truncated ErbB2 receptor (p95^{ErbB2}) is regulated by heregulin through heterodimer
formation with ErbB3 yet remains sensitive to the dual EGFR/ErbB2 kinase inhibitor GW572016." *Oncogene* 2004; 23:646–
53.

3
Product Portfolio Planning and Management in the Pharmaceutical Industry

Alan Woodall

Contents

Industry Immersion Learning. Real-Life Industry Case-Studies in Biotechnology and Business
L. Borbye, M. Stocum, A. Woodall, C. Pearce, E. Sale, W. Barrett, L. Clontz, A. Peterson, J. Shaeffer
Copyright © 2009 WILEY-VCH Verlag GmbH & Co. KGaA, Weinheim
ISBN: 978-3-527-32408-8

3.1
Mission

This case will familiarize students with basic concepts and methods of the planning and management of a pharmaceutical drug portfolio.

3.2
Goals

This case will

- demonstrate the application of basic financial assessment concepts to new pharmaceutical products;
- demonstrate methods to integrate risk and financial assessment in evaluating new pharmaceutical products; and
- introduce methods to continuously evaluate the impact of changing commercial potential and product profile throughout the product development lifecycle.

3.3
Predicted Learning Outcomes

Students will acquire

- an ability to understand and participate in the process of portfolio planning and management in the pharmaceutical/biotechnology industry, including candidate selection and licensing evaluations; and
- an ability to compile and analyze competitive data for pharmaceutical/biotechnology industry products.

3.4
Introduction

The discovery, evaluation, development, approval, and marketing of new products are the lifeblood of the pharmaceutical, biotechnology, and medical device industries. Products discovered and developed by pharmaceutical, biotech, and device companies in the 1970s and 1980s fueled dramatic growth in both the number and size of pharmaceutical companies.[1] The so-called blockbuster drugs (annual sales of more than $ 1 billion) made the industry one of the world's most profitable of the last quarter century.[2]

In order to maintain this growth, much emphasis and money has been given to the discovery, development, and approval of new drug candidates. Discovery of new products is usually driven by identification of a current or future unmet medical need in a patient population or through a technology-based platform that concentrates on identifying a biological process or target receptor(s) or gene(s) and searching for a medical application. Products can come from natural sources, genetic manipulation, or biologically manufactured or chemically synthesized compounds.

Once identified, a potential product is patented and put through a series of preliminary screens designed to determine its likely biological characteristics (toxicology/safety, pharmacology, manufacturing potential, etc.). If the product passes these preliminary screens, a profile of what is known scientifically about the product and its potential market is prepared and used as a basis to consider the product for inclusion in the company's new product portfolio. After selection, the product enters the new product development phase,[3] which typically includes:

1) Congress of the United States, Congressional Budget Office. "Research and Development in the Pharmaceutical Industry." *A CBO Study* October 2006; 39–41. Available at: http://www.cbo.gov/ftpdocs/76xx/doc7615/10-02-DrugR-D.pdf (last accessed 5 September 2008).

2) Berndt, E. "The US pharmaceutical industry: why major growth in times of cost containment?" *Health Affairs* 2001; 20(2):100–114. Available at: http://content.healthaffairs.org/cgi/reprint/20/2/100.pdf (last accessed 5 September 2008).

3) Alliance Pharmaceutical Corporation, Wierenga, D., Eaton, C. Office of Research and Development Pharmaceutical Research and Manufacturers Association, http://www.allp.com/drug_dev.htm.

Preclinical Phase   This phase involves animal testing designed primarily to predict the safety of limited human administration. It typically lasts one to three years, after which an application for human administration is submitted to the appropriate regulatory agency. In the United States, this application is called an Investigational New Drug (IND) application.

Phase I   Following a successful preclinical phase, the product is administered to a limited number of humans (typically 10–75). The human subjects who receive the product are healthy volunteers who are not taking other drugs and do not exhibit signs or symptoms of any disease, including the target disease of the new product. This phase is designed to assess basic human safety, pharmacology, and pharmacokinetics. While phase I studies are often carried out throughout the development of the product, the primary phase I studies last approximately one year.

Phase II   Following successful phase I safety assessment, the product is administered to humans (typically 100–700) who have the target illness. Phase II is designed to (1) determine the appropriate dose regimen (amount, frequency, duration, and timing), (2) characterize the patient populations for whom effectiveness has been demonstrated, (3) determine the initial safety-to-effectiveness profile, and (4) characterize the dose–exposure–response relationships (safety and effectiveness). Phase II typically lasts between one and three years.

Phase III   This phase is designed to generate evidence for safety and effectiveness of the product and involves many patients (more than 1000). It confirms information obtained during phases I and II and follows patients through a defined course of treatment that, if successful, will provide the large body of human safety and effectiveness data necessary for product approval.

Marketing Application   Following a successful phase III, data obtained during all clinical phases of development are combined with nonclinical information on the chemistry, manufacturing, acute and long-term toxicology, and animal pharmacology for submission to the appropriate regulatory agency. In the United States, this application is known as a New Drug Application (NDA).

Phase IV   Following marketing approval, additional clinical and nonclinical studies are conducted to assess the product's safety and effectiveness in large-scale use. These studies may be mandated by the regulatory agency as a condition of approval. Phase IV studies also are conducted as a part of a company's market expansion program and may include efforts to develop additional dosage forms and packaging systems, and to assess the potential for use in new indications or patient populations.

As new information is gained during the process of development of a new product, the product is periodically evaluated to ensure that it remains a viable product and is still compatible with the overall portfolio mission. Only approximately one in 10 new product candidates will eventually reach the market.[4] The process used to select products for inclusion in the portfolio and to periodically confirm their viability and overall fit with company objectives is known as portfolio planning and management.

Prior to the 1990s, the value of new product candidates was most often assessed by company management without the aid of rigorous quantitative systems. Beginning in the early 1990s, several factors dramatically changed the landscape in pharmaceutical new product development and necessitated a more stringent and comprehensive evaluation of new product candidates. First, the cost of developing a new product increased dramatically from an estimated $ 256 million in 1987 for a new chemical entity to approximately $ 802 million in 2001.[5] Second, increasing scientific complexity and regulatory requirements added to the time required to develop new products. Third, with many lucrative markets being satisfied by current products, companies were forced to investigate products in increasingly more risky areas and abandon products that were not the first of their class to be introduced or did not have advantages that could be used to differentiate them from competitors. Fourth, an increasingly competitive environment shrunk the time for which new products were able to maintain market exclusivity.[6] These factors motivated an unprecedented consolidation of pharmaceutical companies, dramatically increased R & D[7] budgets and heightened the need for improved portfolio planning and management techniques, with an increased focus on product candidate selection.

Approaches to portfolio planning and management vary widely within the industry and range from simple qualitative applications to sophisticated simulation models that require years to implement. For small companies with a small number of new product opportunities, the portfolio planning and management process is often largely confined to evaluating new product opportunities, while large companies with many product opportunities spend considerable time both selecting products and optimizing the balance of those currently under development.

Qualitative approaches are typically the focus of a senior management committee that considers information such as medical need, commercial potential, risk, development timing, and strategic product fit to select and prioritize product opportunities. Such approaches are most often used by small and medium-sized companies that do not have a large number of product opportunities and are therefore required to make the most of what is currently available.

---

4) DiMasi, J. "Risks in new drug development: approval success rates for investigational drugs." *Clinical Pharmacology and Therapeutics* 2001; 69(5):297–307. Available at: http://www.nature.com/clpt/journal/v69/n5/pdf/clpt200138a.pdf (last accessed 5 September 2008).

5) DiMasi, J. A., Hansen, R. W., Grabowski, H. G. "The price of innovation: new estimates of drug development costs." *Journal of Health Economics* 2003;22:151–185.

6) Kearney, A.: Too Clever by Half. *The Economist* 20 September 1997, 344:67–68.

7) See Glossary in Appendix B of this chapter.

More sophisticated approaches, often employed by larger companies, apply statistical methods (e.g., Monte Carlo simulation[8]) not only to evaluate new product opportunities but to continuously evaluate products during their development and optimize product development decisions. Enterprise software is also utilized to track costs, predict resource utilization, and summarize information for management (e.g., "dashboards"[13]). These systems require both detailed product development decision networks for each product and highly trained staff to ensure accuracy.

Regardless of the nature of the portfolio planning and management process, all applications provide two basic types of information: financial and scientific/regulatory. The financial information typically consists of cash flow analysis, sensitivity analysis of key variables, and expected and net present value calculations, while the scientific/regulatory information consists of requirements or probability of achieving regulatory approval, meeting an unmet medical need, and the product differentiation necessary for success in the marketplace.

This exercise will attempt to familiarize students with the basic tenets of portfolio planning and management. First, students will be asked to use financial and risk-based approaches to evaluate and prioritize a new product portfolio. Second, students will prepare scientific/regulatory target product profiles suitable for guiding individual product development and evaluating multiple product portfolio opportunities.

3.5
Case Scenario

A newly formed pharmaceutical company is currently evaluating products for inclusion in its new product portfolio. The company is backed by a large private equity concern that has funded the company with sufficient cash reserves to sustain it for four years, at which time it is expected that at least one new product will be marketed.

The strategic intent of the investors is to develop the company into a leadership position in the central nervous system (CNS) market. Company management has determined that it will have to develop at least two products that command a dominant position (more than 30% market share) in order to achieve the stated objectives.

Company management is considering five product opportunities for inclusion in its development portfolio. The challenge is to evaluate all five products and to determine possible scenarios to accomplish the following:

---

8) See, for example, Huber, T. Department of Physics, Gustavus Adolphus College (July 1997), http://physics.gac.edu/~huber/envision/instruct/montecar.htm (last accessed 5 September 2008).

- Evaluate the five product candidates for inclusion in the portfolio.
- Prioritize the five products.
- Balance risk and reward to maximize return on investment.
- Develop a method to periodically measure progress and assess the potential value of each project and the overall portfolio.

3.6
Timeline

Session 1: Case Introduction and Background Information
Session 2: Impact of Risk and Time Value of Money
Session 3: Financial Analysis
Session 4: Target Product Profile Overview
Session 5: Target Product Profile Review
Session 6: Target Product Profile Presentations
Session 7: Automated Portfolio Planning and Management Systems
Session 8: Project Presentations

3.7
Study Plan and Assignments

3.7.1
Session 1

3.7.1.1 **Presentations**
Overview of Portfolio Planning and Management; Assignment 1: Preliminary Prioritization (Presentation #6)

3.7.1.2 **Assignment #1: Preliminary Prioritization**
Preliminary data on the five product opportunities have been received by the company. A summary of the data received is listed in Table 3.1. Review the data and provide

- a priority for development of each product;
- the rationale behind the priorities;
- a list of assumptions that support the priorities; and
- additional data needed to complete a more thorough analysis.

Possible Solutions for Assignment 1
The purpose of this assignment is to encourage students to begin to think about financial possibilities and the information necessary to adequately complete an evaluation. Given the limited data provided, there are no incorrect answers; learning will be provided by discussion amongst the students.

Table 3.1 Product summary.

Product indication	Cost to develop ($ millions)	Time to develop (years)	Annual sales ($ millions)
Alzheimer's disease	1000	4	10600
Anxiety	500	5	1300
Analgesia	525	3	350
Cognitive enhancement	600	6	2500
Nicotine addiction	875	7	15000

Review of the table can begin to narrow down the possibilities. Some obvious observations are:

1. *Priority ranking:*
   - The Alzheimer's disease and nicotine addiction products have huge sales potential and relatively reasonable sales potential/development cost ratios. The anxiety and analgesia products are on the other end of the spectrum, while cognitive enhancement falls somewhere in the middle ground.
   - Of the two products that can be developed within the four-year window of funding established by the company and its investors, the Alzheimer's disease product is the obvious choice. Therefore, the Alzheimer's disease product would receive a higher priority ranking than the analgesia product.
   - Arguments for various rankings could be made based on the relative value placed on time required for development versus sales potential, the most dramatic example being choosing between analgesia with low sales potential but quick time to market and nicotine addiction with high sales potential and a relatively long time to market.
2. *Assumptions:* Within the context of the data provided, the assumptions will likely concern the previously mentioned time to market versus sales potential trade-offs.
3. *Additional data needed:* These data could come from any area of drug development that has a financial impact on the development and marketing of pharmaceutical products. Possible examples include:
   - Manufacturing costs
   - Capital investments
   - Inventory costs
   - Marketing expenses
   - Product launch expenses
   - Inflation (sales and costs)
   - Taxes
   - Costs for post-approval development (new indications, formulations, packaging, etc.)

Read:
Introduction to this chapter

3.7.2
Session 2

3.7.2.1 **Presentations**
Expected Value and Net Present Value; Assignment 2: Risk-Adjusted Prioritiza-
tion; Assignment 2: Possible Solutions – Graphical Formats; Net Present Value;
Assignment 3: Financial Analysis Without Risk (Presentation #7)

3.7.2.2 **Assignment #2: Risk-Adjusted Prioritization**
Company scientists have analyzed scientific data for the five products and estimated the probability of successful development for each product. The probabilities are listed in Table 3.2. Given these new data, discuss how the inclusion of
probability estimates affects the relative value of the products and list rationale
for any changes in priority. How might these results be presented to management in a graphical format?

Possible Solutions for Assignment 2
The purpose of this assignment is to introduce the potential impact of probability weighting and the role of expected value into the exercise. Multiplying the
annual sales by the probability of success yields the expected sales for each
product. These probability weighted values can be used to review and revise the
priorities.

Inclusion of Probability and Rationale for Changes
Review of the data (listed in Table 3.3) suggests the following:

• The relatively low probability of success given to the Alzheimer's disease
  product significantly reduces its expected sales value to below that of the
  cognitive enhancement product.
• The nicotine addiction product remains the most valuable product.
• Of the two products that can be developed within the company's four-year
  time frame, the Alzheimer's disease product retains its higher value.

Table 3.2 Products' probabilities of success.

Product indication	Probability of success (%)
Alzheimer's disease	10
Anxiety	20
Analgesia	60
Cognitive enhancement	80
Nicotine addiction	50

Table 3.3 Product data.

	Alzheimer's disease	Anxiety	Analgesia	Cognitive enhancement	Nicotine addiction
Annual sales ($ millions)	10600	1300	350	2500	15000
Probability of success (%)	10	20	60	80	50
Expected sales ($ millions)	1060	260	210	2000	7500
Cost to develop ($ millions)	1000	500	525	600	875
Time to develop (years)	4	5	3	6	7

3.7.2.4 Assignment #3: Financial Analysis Without Risk

After reviewing the analysis produced in Assignment 1, management requested a more comprehensive analysis that values each product from a current value (i.e., net present value [9]) point of view. Include the additional data required to complete a comprehensive analysis identified in Assignment 1 and provide a comprehensive analysis of all five product opportunities. Assume all products will be successfully developed (i.e., probability of success = 100%) and will remain viable in the market until patent expiration in 10 years. How does the addition of these data change the priorities?

Possible Solutions for Assignment 3

The purpose of this assignment is to consider the effects of financial variables that significantly affect most pharmaceutical products.

Using the assumptions listed in the possible solutions for Assignment 1, Table 3.4 summarizes the analysis assuming successful development of each product. To simplify the exercise, amounts for the following variables were assumed to be similar for all five products and are not included in the analysis.

- Post-approval development
- Capital investments
- Inventory costs
- Product launch expenses
- Inflation in sales and costs

Marketing and manufacturing expenses (cost of sales) were assumed to be 15% each, and taxes were assumed to be 30% of sales. A 15% discount rate was used for net present value calculations, and sales projections were assumed to continue only until patent expiration. The method used for net present value calculations is listed in Appendix A to this chapter.

Changes Resulting from Additional Data

While the overall priorities have not changed from Assignment 1, the time value of money (net present value) has reduced the value of the nicotine addiction product to a lower value than that of the Alzheimer's disease product.

9) See Glossary in Appendix B of this chapter.

Table 3.4 Product analysis summary.

	Alzheimer's disease	Anxiety	Analgesia	Cognitive enhancement	Nicotine addiction
Annual sales ($ millions)	10600	1300	350	2500	15000
Probability of success (%)	100	100	100	100	100
Expected annual sales ($ millions)	10600	1300	350	2500	15000
Cost of sales ($ millions)	3180	390	105	750	4500
Earnings ($ millions)	7420	910	245	1750	10500
Taxes ($ millions)	2226	273	74	525	3150
Earnings after tax ($ millions)	5194	637	171	1225	7350
Development costs ($ millions)	1000	500	525	600	875
Development time (Years)	4	5	3	6	7
Net present value ($ millions)	10525	726	68	1134	5789

3.7.3
Session 3

3.7.3.1 Presentations
Possible Solutions for Assignment 3; Assignment 4: Financial Analysis With Risk (Presentation #8)

3.7.3.2 Assignment #4: Financial Analysis With Risk
After reviewing the analysis provided in Assignment 3, management has requested that the analysis be risk adjusted. Using the probability of success provided in Assignment 2, repeat the analysis allowing for risk. How does the addition of risk alter the analysis? Which factors would persuade management to
- use the risk-adjusted analysis as a basis for selecting or prioritizing products?
- use the non-risk-adjusted analysis as a basis for selecting or prioritizing products?
- ignore both analyses?

Possible Solutions for Assignment 4
Table 3.5 repeats the analysis conducted in Assignment 3 except that risk data are used to adjust the sales values. All other calculations are the same.

Table 3.5 Risk data used to adjust sales values.

	Alzheimer's disease	Anxiety	Analgesia	Cognitive enhancement	Nicotine addiction
Annual sales ($ millions)	10600	1300	350	2500	15000
Probability of success (%)	10	20	60	80	50
Expected annual sales ($ millions)	1060	260	210	2000	7500
Cost of sales ($ millions)	318	78	63	600	2250
Earnings ($ millions)	742	182	147	1400	5250
Taxes ($ millions)	223	55	44	420	1575
Earnings after tax ($ millions)	519	127	103	980	3675
Development costs ($ millions)	1000	500	525	600	875
Development time (years)	4	5	3	6	7
Net present value ($ millions)	409	−123	−118	831	2634

Changes Resulting from Inclusion of Risk
The high sales and probability of success for the nicotine addiction product produce a significantly higher value than the other four products. Due to their negative net present value, the analgesia and anxiety products are theoretically not worthy of development. The cognitive enhancement product becomes marginally more valuable than the Alzheimer's disease product.

Factors That Would Persuade Management to Use the Non-Risk-Adjusted Analysis
An argument can be made that a product will either fail or be successful. Risk-adjusted assessments are most accurate when applied to large populations and may not be appropriate for evaluation of only five products.

Factors That Would Persuade Management to Use the Risk-Adjusted Analysis
The high failure rate (around 90%) of pharmaceutical products and the long development horizon will produce more value over the long run.

Factors That Would Persuade Management to Ignore Both Analyses
A "loss leader" approach to developing a market presence in advance of an important product or simple altruism and desire for the betterment of mankind are possible explanations.

3.7.4
Session 4

3.7.4.1 **Presentations**
Possible Solutions for Assignment 4 and Target Product Profile; Target Product Profile; Assignment 5: Target Product Profile (Presentation #9)

3.7.4.2 Assignment #5: Target Product Profile (Introduction)

Now that the company understands the financial implications for the development of each product candidate, it wishes to ensure that each product is scientifically evaluated and developed in the best manner possible. To accomplish this, management has requested that a Target Product Profile[10] (TPP) be prepared for each product. The TPP is an inventory of key development attributes that are required for successful development, approval, and marketing of a product. While it is primarily a scientific document, the TPP is developed using techniques similar to a SWOT (strengths, weaknesses, opportunities, and threats) analysis.[11]

Management has requested that the TPP compare each of the five development candidates to its likely competitor and provide an estimate of the minimum attributes necessary in order to achieve the desired market share. Names of the competitor products are identified in Table 3.6. A brief outline of a typical TPP is contained in Table 3.7.

Select one of the five product opportunities and compare it with information on the likely competitor product. Identify the TPP elements that offer the best opportunity for product differentiation.

Possible Solutions for Assignment 5

The purpose of Assignment 5 is to suggest the TPP as a primary document to evaluate and manage development of a pharmaceutical product. While many solutions exist, the important learning will come from exploring the competitor information to identify areas of potential product differentiation. Following are some of the obvious opportunities for differentiating each product.

Alzheimer's disease product:
- A product that alters the course of the disease rather than treating only the symptoms.
- A product with improved efficacy; clinical trials conducted on Aricept demonstrate marginal efficacy (ADAS-cog and CIBC Plus scales[12]).

Anxiety product:
- Zoloft® is indicated for social anxiety disorder, not generalized anxiety disorder. Therefore, it is presumed that all Zoloft® prescriptions for generalized anxiety disorder are "off label." An FDA-approved indication for generalized anxiety disorder for the product would provide a significant differentiation opportunity.
- A non-SSRI, or a product without the suicide liability.

10) Gibson, M. "Pharmaceutical Preformulation and Formulation: A Practical Guide from Candidate Drug Selection to Commercial Dosage Formulation", Informa Healthcare 2001, pages 157–160

11) For SWOT analysis, see http://www.mindtools.com/pages/article/newTMC_05.htm (last accessed 23 September 2008).

12) For the ADAS-cog use Galaska, D., Bennett, D., Sano, M., Ernesto, C., Thomas, R.,

Grundman, M., Ferris, S. An inventory to assess activities of daily living for clinical trials in Alzheimer's disease. The Alzheimer's Disease Cooperative Study. Alzheimer Dis Assoc Disord 1997, 11(Suppl 2):S33–39. For the CIBC Plus scale use Knopman, D. S., Knapp, M. J., Gracon, S. I., Davis, C. S. The Clinician Interview-Based Impression (CBI): a clinician's global rating scale in Alzheimer's disease. Neurology 1994, 44:2315–2321

Table 3.6 Sources of product information.

Alzheimer's	Aricept® (donepezil) – Physicians Desk Reference 62nd edition, 2007, ISBN: 1-56363-660-3, Thompson Healthcare Inc. Montvale NJ, Pages 1075–1079
Anxiety	Zoloft® (sertraline) – Physicians Desk Reference 62nd edition, 2007, ISBN: 1-56363-660-3, Thompson Healthcare Inc. Montvale NJ, Pages 2576–2584
Analgesia	OxyContin® (oxycodone) – Physicians Desk Reference 62nd edition, 2007, ISBN: 1-56363-660-3, Thompson Healthcare Inc. Montvale NJ, Pages 2680–2685
Cognitive enhancement	Provigil® (modafinil) – Physicians Desk Reference 62nd edition, 2007, ISBN: 1-56363-660-3, Thompson Healthcare Inc. Montvale NJ, Pages 3466–3471
Nicotine addiction	Nicorette® (nicotine gum) http://www.theodora.com/drugs/nicorette_gum_glaxosmithcline_consumer.html

All URLs last accessed 23 September 2008.

Table 3.7 Outline of a typical target product profile.

TPP element	Minimum required	Maximum possible	Target
Regulatory hurdles			
Overall efficacy			
Onset of action			
Duration of effectiveness			
Overall safety			
Adverse reactions			
Contraindications			
Drug–drug interactions			
Dosage form			
Dose duration			
Dose amount			
Dose frequency			
Route(s) of administration			
Absorption profile			
Distribution profile			
Metabolism profile			
Elimination profile			
Use in special populations			
Abuse liability			
Cost of raw drug			
Cost of finished product			
Shelf life			

Analgesia product:
- A product that avoids known opioid safety issues (respiratory suppression, abuse potential, etc.).
- A product that provides long duration of action without overdose or abuse potential (e.g., from crushed/broken tablets).

Cognitive enhancement product:
- The competitor product is a CNS stimulant approved for the treatment of conditions that cause excessive sleepiness (e.g., narcolepsy, sleep apnea, etc.). All uses of the marketed product for cognitive enhancement are, therefore, off label. An FDA-approved indication for cognitive enhancement would result in near 100% market share.

Nicotine addiction product:
- A non-nicotine product.
- A product that more effectively reduces the urge to resume smoking.
- A product that requires less frequent dosing.

3.7.5
Session 5

Review and work on Assignment 5

3.7.6
Session 6

Review and work on Assignment 5

3.7.7
Session 7

3.7.7.1 **Presentations**
Possible Solutions for Assignment 5 (Presentation #10)

3.7.8
Session 8

3.7.8.1 **Presentations**
Automated Portfolio Planning and Management Systems; Assignment 6 Example Calculations (Presentation #11)

3.7.8.2 **Assignment #6: Automated Portfolio Planning and Management Systems**
Companies with a large number of products in their portfolio, many of which can be at different stages of development, frequently employ computerized enterprise systems to assist with managing the large amount of information required to monitor multiple projects. For portfolio management purposes, enterprise systems compile billing rate information from human resource compensa-

tion records, work effort from employee time reporting/card records, and project status information from project management systems into a single database which is used to produce a variety of management reporting tools. Graphical displays of project status information (budget, costs, timelines, etc.), typically referred to as dashboards,[13] are a common output of this type of system.

For portfolio planning purposes, automated systems are most often utilized to simulate potential development scenarios and predict expected costs for projects based on a range of possible decisions. After deciding on a development candidate and completing a target product profile, company project managers prepare a detailed project network that outlines all the major decisions required for successful project completion. The decision tree outlined in Figure 3.1 is typical of the many decisions contained in a project network. The decision tree in Figure 3.1 outlines two activities (formulation development and formulation scale-up) that are required prior to producing large-scale lots of a product formulation. Each of the two activities has a potential for succeeding or failing. If the activity fails the entire project will be terminated. Automated portfolio planning systems will utilize all project decisions in the detailed project network and simulate project cost and probability for success.

Review the decision tree in Figure 3.1 and calculate:

- the expected costs for completing formulation development
- the expected costs for completing formulation scale-up
- the expected costs for reaching the large-scale production activity
- the probability of reaching the large-scale production activity

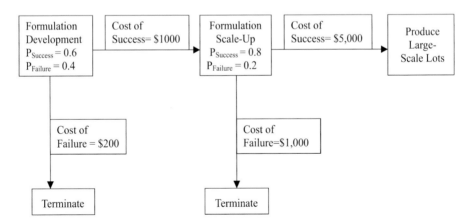

Fig. 3.1 Typical formulation development/scale-up decision tree.

13) See www.bea.com/content/news_events/
white_papers/BEA_PfizerPharmaceuticals_cs.pdf
(last accessed 23 September 2008).

Possible Solutions for Assignment 6

Because it is not known if either formulation development or formulation scale-up will be successful or result in project termination, portfolio planning must consider the cost of both possibilities for each activity. By calculating the expected costs of success and failure for each activity an expected cost can be calculated for the entire decision network. Because project networks contain many (often more than 1000) decisions and large company portfolios contain multiple (often more than 25) projects, this approach is a useful method of estimating overall portfolio success and cost.

Expected cost to complete formulation development =
$(P_{Success} \times$ Cost of success$) + (P_{Failure} \times$ Cost of failure$)$
(where P is probability),

or

$0.6 \times \$ \; 1000 = \$ \; 600$
$0.4 \times \$ \;\; 200 = \$ \;\; 80$
$ \overline{\$ \; 680}$

Expected cost to complete formulation scale-up =
$P_{Success} \times$ Cost of success $+ P_{Failure} \times$ Cost of failure $\times$ Formulation development $P_{Success}$

or

$0.8 \times \$ \; 5000 \times 0.6 = \$ \; 2400$
$0.2 \times \$ \; 1000 \times 0.6 = \$ \;\; 120$
$ \overline{\$ \; 2520}$

Expected cost of reaching large-scale production activity =
Expected cost of formulation development + Expected cost of formulation scale-up

or

$\$ \; 680 + \$ \; 2520 = \$ \; 3200$

Probability of reaching large-scale production activity =
Formulation development $P_{Success} \times$ Formulation scale-up $P_{Success}$

or

$0.6 \times 0.8 = 0.48$

Appendix A: Method for Net Present Value Calculations

Net present values (NPVs) for all products were calculated using the following formula:

t = time of the cash flow (in years)

n = total time of the project

r = discount rate

C_t = net cash flow (the amount of cash) at time t.

C_0 = Initial investment

   (capital outlay at the beginning of the investment time, $t=0$)

As an example, calculations for the Alzheimer's product in Assignment 3 are:

- Total time $= 10$ years (4 years development + 6 years sales)
- Discount rate $(r) = 15$ percent
- Net cash flow $(C_t) = -\$ 250\,000$ (annual development cost) for $t=1$–4 and $+\$ 5\,194\,000$ (annual after-tax earnings) for $t=5$–10
- Initial investment $(C_0) = 0$

The table below lists present value calculations for each year of the project using the above formula. By summing the present values for all 10 years ($\$ 11\,238\,638$–$\$ 713\,742$), the NPV ($\$ 10\,524\,896$) is determined.

Table A.1 Net present value calculations for each year for the Alzheimer product in Assignment 3.

Year (t)	Present value (thousands)
1	$-\$ 250\,000/1.15 = -\$ 217\,391$
2	$-\$ 250\,000/1.15^2 = -\$ 189\,036$
3	$-\$ 250\,000/1.15^3 = -\$ 164\,376$
4	$-\$ 250\,000/1.15^4 = -\$ 142\,939$
5	$\$ 5\,194\,000/1.15^5 = \$ 2\,582\,281$
6	$\$ 5\,194\,000/1.15^6 = \$ 2\,245\,471$
7	$\$ 5\,194\,000/1.15^7 = \$ 1\,952\,632$
8	$\$ 5\,194\,000/1.15^8 = \$ 1\,697\,941$
9	$\$ 5\,194\,000/1.15^9 = \$ 1\,476\,449$
10	$\$ 5\,194\,000/1.15^{10} = \$ 1\,283\,864$
NPV	$\$ 10\,524\,896$

Appendix B: Glossary of Abbreviations and Terms

Absorption profile – A description or measurement of how a drug is absorbed into the body.

Abuse liability – The potential for patients to become addicted or to use the product inappropriately and the resulting impact on development and marketing.

ADME – absorption, distribution, metabolism, and excretion.

Adverse reaction – A harmful or unexpected event associated with the use of a drug product.

ANDA (Abbreviated New Drug Application) – A document submitted to the US Food and Drug Administration requesting approval to market a generic version of a currently marketed drug with an expired patent.

BLA (Biological License Application) – A document submitted to the US Food and Drug Administration requesting approval to market a biological product (virus, therapeutic serum, toxin, antitoxin, vaccine, blood, blood component or derivative, allergenic product, or analogous product applicable to the prevention, treatment, or cure of diseases or injuries of humans).

CFR – Code of Federal Regulations.

CMC – Chemistry, manufacturing, and controls.

CNS – Central nervous system.

Contraindication – A condition for which a drug product should *not* be used.

FIH – First in humans.

Distribution profile – An explanation or measurement of how a drug is spread throughout the body.

Dosage form – Completed forms of a pharmaceutical preparation.

Dose amount – The quantity of a drug product taken.

Dose duration – The length of time for which a drug product is taken.

Dose frequency – How often a drug product is taken.

Drug–drug interaction – The pharmacological result, either desirable or undesirable, of drugs interacting among themselves.

Duration of effectiveness – The length of time for which a product successfully treats the target condition.

Elimination profile – The process or measurement of how the body rids itself of a drug.

IND (Investigational New Drug) – A document submitted to the US Food and Drug Administration requesting permission to administer an investigational drug to humans.

MAA (Marketing Authorization Application) – European Union equivalent of an NDA.

Metabolic profile – The physiological or chemical disposition of a drug within the body.

NDA (New Drug Application) – A document submitted to the US Food and Drug Administration requesting approval to market an investigational drug product.

NPV (Net Present Value) – A discounting technique that measures the value of the present value cash inflows minus cash outflows, usually used to measure profitability of a project.

Onset of action – The time required for a drug to exhibit an effect.

Overall efficacy – The degree to which a product treats its target population, often expressed as the percentage of the target population that is successfully treated.

Overall safety – The cumulative number of adverse reactions and side effects experienced during treatment, a broad and often subjective measure.

Phase I – The first trials in humans, usually designed to evaluate safety, tolerance, and pharmacokinetics.

Phase II – Clinical studies involving patients (i.e., humans with the target disease), usually designed to assess dose regimens.

Phase III – Large clinical trials, usually designed to assess efficacy and drug-related effects.

Phase IV – Studies performed after a drug has been approved for marketing.

Preclinical development – Studies conducted on a potential drug prior to administration to humans.

Regulatory hurdle – Expectation or requirement for product approval dictated by a regulatory authority.

R & D – Research and development.

ROI – Return on investment.

Route of administration – The method or site used to deliver a drug into the body.

Shelf life – The approved length of time a product is allowed to remain in the distribution system.

Special populations – Subsets of patients who are underrepresented, difficult to treat, overly sensitive to treatment, etc.; typically defined by the drug manufacturer or regulatory agency.

SSRI – Selective serotonin reuptake inhibitor.

TPP (Target Product Profile) – A document indicating the attributes necessary and/or desired for successful development of a drug product.

4
Entrepreneurship: Establishing a New Biotechnology Venture

Cedric Pearce

Contents

Industry Immersion Learning. Real-Life Industry Case-Studies in Biotechnology and Business
L. Borbye, M. Stocum, A. Woodall, C. Pearce, E. Sale, W. Barrett, L. Clontz, A. Peterson, J. Shaeffer
Copyright © 2009 WILEY-VCH Verlag GmbH & Co. KGaA, Weinheim
ISBN: 978-3-527-32408-8

4.1
Mission

The mission is for students to collaboratively develop a business proposition into a business plan and presentation for potential investors.

4.2
Goals

The goal for science students with limited business background is to understand how science and technology ideas may be used as a basis for a company. Through rle playing, students will develop an understanding of the commercial and early-stage work environment as well an appreciation of how to start and run their own businesses. One of the most important aspects of this exercise is that the students should be able to work together as a team and try to generate something which is greater than the sum of the collected contributions.

4.3
Predicted Learning Outcomes

Students will

- learn to explore independent business ideas and how to take responsibility for their own professional future;
- achieve an understanding of the organizations for which they might at some stage in their careers be involved.

4.4
Introduction

Question: What is the quickest way to make a small fortune in biotechnology?
Answer: Start with a large one.

Do you want to be a career professional or an entrepreneur? This is one of the questions that the student will have to answer when entering the workforce.

The traditional route for a science graduate is to become a bench-working scientist, a career professional within an established company, and to fall into the multilayered hierarchical structure of that company. Given the dedication and hard work it takes to develop the skills needed to be a scientist, the drive to preserve this educational investment is often expressed through continued involvement with the technical details of a particular field.

There are few rules about starting your own company. What is presented here are opinions which have helped the author in founding and running three companies, mixed in with observations and advice from various literature sources and courses.

4.4.1
Characteristics of an Entrepreneur

Of the various definitions of an entrepreneur, we shall assume the term refers to a person who takes charge and organizes, manages, and assumes responsibility for a business or other enterprise.

Entrepreneurs are often referred to as risk takers; this does not mean they are reckless, but that they are willing to undertake something the outcome of which is uncertain. The result may be desirable, with satisfying rewards, or it may be unsuccessful and have other downsides. In the case of the entrepreneur, risk is used to describe the uncertainty associated with the outcome of a business activity. Gambling or betting on the outcome of a horse race, for example, also has an uncertain outcome, although the result is not (normally!) within the bettor's control. Usually in entrepreneurship, although there is an element of luck, the overall success of a set of actions is strongly influenced by the entrepreneur's actions; thus the entrepreneur carefully evaluates and then manages the risk.

The entrepreneur undertakes a series of actions with a certain goal in mind, but with no guarantee that this will be achieved. There are many good reasons why a person might want to pursue such activity. Without taking risks, progress in any field is going to be slow at best. There are a number of publications that address risk and reward; in particular, two books come to mind: *Against the Gods: The Remarkable Story of Risk,* by Peter L. Bernstein, and *Pioneering Research: A Risk worth Taking,* by Donald W. Braben.

Some of the most exciting opportunities for scientists are in early-stage companies, exploring commercialization of cutting-edge science and technology. Knowledge of how entrepreneurs approach issues and how start-up companies are organized and operate is essential if the scientist is to prosper in this environment. Entrepreneurs recognize that problems are opportunities and are willing to invest time and effort in solving the problem, with the expectation that value will be generated. Often, entrepreneurs would rather solve a problem through their own business than solve the same problem while being employed in another person's business.

An entrepreneur will recognize an unmet need and a corresponding solution, or at least have an idea for a solution. This is only one part of the puzzle; it also needs to be determined that there is a market for that solution; the risks need to be evaluated, which might involve some education; a plan needs to be developed, which addresses the resources that need to be marshaled to address the problem; and, finally, a team needs to be assembled to address the issue. Flexibility is a needed characteristic. The ability to thrive in a complicated and sometimes chaotic, potentially stressful environment is useful and possibly even essential.

The financial situation of the entrepreneur is important in an evaluation of the prospects of success. Can the person afford to be without a day job with a regular income stream? A person who is indebted, whether from student loans, credit cards, mortgages, or other bank loans, which will require servicing, takes risks with his or her income. On the other hand, being able to be without a regular salary for as little as three months is a good starting position from which to undertake an entrepreneurial adventure. At times, being an entrepreneur can be lonely and challenging; self-confidence and optimism help an individual get through tough times when nothing seems to go right or when bucking the general trend to try something different.

Motivation to work on your own projects can come from any of a variety sources. Financial motivation is often used; independence is important to some, and for many others the will and need to address more or less ignored problems simply for the good that it will do if successful is sufficient reward in itself.

4.4.2
Twenty Questions to Determine How Entrepreneurial You Are

The following brief exercise illustrates some common entrepreneurial features. It is meant to assist students to evaluate whether the entrepreneurial situation is one they would be happy with or want to avoid. The goal is to encourage reflection on the possible advantages and disadvantages of corporate careers versus self-employment. There are no wrong or right answers, only a promotion of self-insight. For the person more suited to a corporate career, being in an entrepreneurial situation may be no picnic, and vice versa.

1. Are you able to identify a problem and fix it without being told to?
2. Is the glass half full, or half empty?
3. Where others see problems do you see opportunities?
4. Do you like to be a leader?
5. Do you dislike difficult challenges?
6. Do you believe that you can make things happen?
7. Do you enjoy the big picture rather than the technical details?
8. Would you rather fail trying something difficult than succeed doing something straightforward?
9. Would you be put off doing a difficult task until another time if something more interesting but less urgent comes up?
10. Do you take full responsibility for your actions?

11. Would you rather work by yourself than with other people?
12. Are you disorganized?
13. Would you rather come to a decision after carefully considering all of the problems that might occur?
14. Do you enjoy trying new things?
15. Do you follow the mainstream opinion?
16. Are you able to maintain careful records in laboratory classes?
17. In the laboratory, do you like to try different experiments even though you know they will not all work as planned?
18. Do you like to work hard to achieve a high level in whatever you are doing?
19. Would you rather be rewarded immediately than have to wait, even if by waiting your reward is greater?
20. Do you agree that uncertainty, confusion, and chaos can belong in any work place?

Score one point for each of the following questions to which you answered "Yes": 1, 3, 4, 6, 7, 8, 10, 14, 15, 16, 17, 18, 20

Score one point for each of the following questions to which you answered "No": 5, 9, 11, 12, 13, 19Score one point if the glass in question 2 is half full.

If your answer to question 8 is, "Yes, but I won't fail", give yourself a bonus point.

The higher your score, the more entrepreneurial you tend to be.

4.4.3
From Idea to Concept Evaluation

It is not uncommon for a new scientific discovery to have potential product applications, which can be sold, licensed, or otherwise leveraged to the advantage of the entrepreneur. The idea should directly address a need. Convincing oneself that a real need exists should be completed before progressing through the process of setting up a business and all that entails.

In the pharmaceutical industry, a new antibiotic that kills drug-resistant pathogenic bacteria might be a very desirable thing to develop and market, whereas another cholesterol-lowering agent might not be such an obvious unmet medical need. The potential market for a new platform technology, i.e., a broadly enabling technique with potentially multiple applications, for example robotics and automation in high throughput screening, might be much harder to determine, especially if it will be considered in relationship to a portfolio of other technologies. For example, a new molecular biology technique for determining the effectiveness of a novel medicine might seem a reasonable fit with current animal testing, but, once developed and approved by the Food and Drug Administration (FDA), might be made redundant by subsequently developed technologies.

A company built around a single idea or compound is very risky, while one with a broadly applicable technology or range of compounds, for example, will have less risk. On the other hand, it needs to be kept in mind that focus is important, and the business portfolio should be balanced. Breakthrough or revolutionary technical ideas present a unique problem in relation to ideas that are evolutionary because customers for a truly novel product often are not able to determine its value for their situation.

Because a lot of energy and resources are going to be employed in developing the idea into a business, the entrepreneur needs to be as certain as possible that there is a need for that product *before starting work*. Many technical startups fall into the trap of "If I build it, they will buy it", when in reality there may be little interest in the product. Market research will enable the entrepreneur to test the validity of the business plan and provide basic information about customer and industry needs. A scientist who is trained to be skeptical in every aspect of his or her professional life needs to also consider market research data to be valuable and capable of preventing a serious mistake, i.e., the spending of a lot of time developing something that is of interest only to a limited section of the potential market. With a little forethought and research, the same idea could maybe have been developed to address a particular need, a process sometimes referred to as the "pain" (see "Resources", *A Good Hard Kick in the Ass: Basic Training for Entrepreneurs*, by Rob Adams).

Some of the basic questions to ask are:

- Who is the customer?
- Is there a true need for this product?
- How long will there be a need for the product?
- What size is the market?
- What is the status of the competition?
- Is this an evolving market, and is there a future need for a refined product?
- How much can be charged for the service or product?
- Does this represent a profitable business?
- Does this represent a growth business?

A questionnaire that addresses potential customers can be helpful. To evaluate if the questionnaire is going to generate the needed information, an exploratory project addressing data gathering and involving a small subset of targeted individuals or companies may be appropriate.

The Customer. The unmet need will help identify the customer; however, the goal at this stage of the business is to obtain data addressing the market potential of the business product, and to do this, customer involvement is important. Customers can be identified through knowledge of the appropriate industry, and an experienced business development professional can provide guidance. Engaging potential customers will enable the start-up entrepreneur to determine the exact requirements of the product. Response to a set of predetermined questions will provide qualitative information. Interaction with customers can be direct, over the phone, or via the Internet or traditional mail.

The Need. The need can be determined by simple questions, i.e., "If we make this, will you use it?" More desirable is the need for a quantitative response; the entrepreneur wants to determine the exact requirements that will satisfy the customer. For a technical device such as a new type of assay, there may be competition from an alternative source, which may act as a barrier to implementation. However, the existence of an alternative source may also provide knowledge concerning what the product needs to exceed in terms of capabilities, which in turn will provide specifications for the new product. For a new type of therapeutic, there may be no competition, but customers could have a preferred set of desires for the product, which would lift it from acceptable to highly desirable.

The Market. In order to assess the market potential, both the number of customers there are and how much they are willing to pay for the product should be determined. When customers have been identified in the first stage of the market analysis, this group will provide feedback concerning the value of the product. By projecting the number of potential customers to the overall population being targeted, it may be possible to estimate the total market size.

The Competition. Information concerning competitors can be obtained from the Internet, the library, or professional organizations, or more likely a mixture of all three. Attending a relevant conference could provide a gold mine's worth of data. Searching patent databases may be particularly helpful (see Chapter 6). If a new product is being developed, it may be difficult to identify the competition. Knowing which companies are working in a particular area may help. In the absence of any firm data, the safest position is to assume there is competition.

The Future. It is important to know if second-generation products are possible and desirable. The customer base contacted initially could be contacted to either test the entrepreneur's future ideas and/or obtain ideas for likely products.

Some of the information needed may be available either publicly or through a commercial organization. In the field of medicine, this information may be freely available and of very high quality. The US National Institutes of Health will frequently provide information concerning a particular unmet medical need. For example, the Alliance for Tuberculosis (TB) Drug Development has produced a detailed document addressing both the need and the potential market for new tuberculosis medicines. It would be appropriate to use this data in a business plan, referring to the source of the data. In this way the value of any new antibiotic for TB can be justified, and any potential investor reasonably reassured of the future of the product. Similarly good, up-to-date information can be found on the various Web sites for foundations and organizations as well as potential competitors for a variety of disease treatments. Other sources that could be employed include libraries, professional organizations and societies, and industry meetings and conferences where there is an opportunity to meet and discuss ideas with colleagues; obviously discretion is needed when discussing information that needs to be kept proprietary.

If a person has a great idea for a business, there is a good chance that someone, somewhere has the same idea or would like to copy the idea and exploit it for his or her own gain. All business is competition, and the entrepreneur needs to protect intellectual property (IP) to the degree that the law allows.

There are different ways to protect an idea and these are described in detail in Chapters 5 and 6.

4.4.4
Assessing Opportunity and Writing a Business Plan

There is a huge amount of entrepreneurial activity within the biotechnology field, and there are many opportunities to join early-stage companies. How these early-stage businesses operate, what to expect, what the culture might be like, how it might differ from an established organization, and how the new employee will fit and prosper are important issues.

Once an idea has been validated as an unmet need, and the founders convinced that it represents a solid business idea, the next step is to draft a business plan. For this purpose, one may consider answering the following questions:

- What type of organization will serve both the entrepreneur (you) and the customer best? Do you anticipate that this will be a growth company that will expand over the course of three to five years and have hundreds of employees or a steady revenue-generating company that will provide an income for the founders and the current employees?
- How will the start-up be funded? Will you use your own resources, a bank or similar loan, or funding from friends and family or angel or venture investors to start and grow the business? Are there government programs that can help, such as Small Business Innovation Research grants? Is local support available for the new business? Is there some formal help? In the real world, a scientist who starts and runs a business should consider the involvement of professionals and consultants. There are a number of organizations that offer help to the business neophyte. A useful consultant should be familiar with the area in which the business operates. For example, if the technology being developed is a new medical device, then the help of a business consultant who is familiar with medicine and medical devices and knows something about how the FDA operates should be a goal. Also, a university will frequently have a technology transfer department, and individuals within this group might be available to either help directly or recommend other sources of assistance. As an exercise, the university student should investigate his or her institution's support for new companies.
- Do you want or need a partner? There are advantages and disadvantages both to having a partner and to working alone. If you want to develop a growth organization, you will need to start hiring staff, in which case an early partner might be a good addition.
- At what point will the company generate revenues/become profitable? This depends on the type of company. A service company probably will generate revenues early in the company's life, whereas an organization involved with developing new medicines will almost certainly take longer to generate profit.

A business plan is a good way to share an idea with others. The process of writing the plan provides an opportunity for the founders to evaluate their idea. A business plan normally consists of the following:

- Title page, including authors and date
- Table of contents
- Executive summary
- Mission statement
- Management team and organization
- Technical description of the business idea
- Status of technology and competition
- Business model
- Financial aspects

The business plan does not have to be lengthy and is often an evolving document. A number of professional organizations are available to help develop the business plan, as well as software packages that make the process more or less straightforward. The business plan often will form the basis for various company presentations.

Company Name. The name of the business should be striking and convey some meaning. However, naming any business is quite difficult, and there are many interesting and scholarly articles written about this subject. It is difficult to know what makes a good name, but there are some general guidelines to follow. Small businesses are often named after the founders; for example, Pfizer and Merck are family names (they started small and presumably did not want to change their names once they had grown). Another approach is to name the business according to the technology being employed, for example, Calgene, indicating a molecular biology company located in California, or Genentech, a contraction of gene and technology. It is generally considered a poor idea to use initials, but there are some great exceptions.

Executive Summary. The executive summary is often written after a draft of the business plan has been developed. Two or three paragraphs will usually describe the essence of the business plan, illustrate the unmet need that is being addressed, how the technology will solve the problem, and how revenues will be generated. The management team would also be featured in the summary.

Mission Statement. The mission statement is placed first, before the remainder of the business plan. It is a short, concise description of what the company wants to do, and is usually one to two sentences long. The mission statement can sometimes be accompanied by specific goals statements. Such goals can briefly describe products and ways to obtain products.

Management Team and Organization. Assembling a good management team is almost always critical to business success, although some ideas are so strong that a business can be developed despite poor management. Investors frequently are attracted by the management team before they learn about the idea, and probably more emphasis is put on the team than the idea. (This is a chicken-and-egg situation, because it takes a good business idea to attract an experienced business team in the first place, and the experienced team will also develop the idea further. See below, "Organizing the Venture".) Initially, the following officers will need to be identified: chief executive, science and technology leader, finance officer, legal advisor, and members of the board of advisors or directors. In the business plan, these officers and their reporting structure are often visualized in organizational charts.

Technical Description of the Business Idea. The technical description of the plan needs to provide an overview and should not be overly complex. A balance is required between providing sufficient detail and writing a scientific treatise. Additional details can be explained in meetings with potential investors and other interested parties. A brief review of the field may be helpful.

Status of Technology and Competition. Competing technology and other companies working in the same field need to be fully disclosed in the business plan. This is important because other companies engaged in research and development of similar technologies may limit the freedom to operate through possession of intellectual property. In addition, such companies could also be evaluated as potential collaborators or licensees if the technology described in the business plan is of interest to them.

Business Model. A description of how the business will generate revenues should be included. A service-oriented company may be able to generate revenues quickly. Early income can be viewed as potential profit for the founders or may be reinvested in the technology and other aspects of the business. In the case of technology companies, often the technology itself might need to be adjusted several times before it can be marketed, and any potential income stream deferred until such time as a product is developed. This is often the case with an early invention obtained from a university, and investment will be required to cover the costs of operating the business until a product is finally developed. If a product needs federal approval prior to marketing to the public, such as a new drug or medical device, this may take a number of years (see Chapter 3). A new drug, for example, has to be shown to be both safe and to have the desired effect; there are strict FDA rules about this process, which involves phase I, II, and III studies, in which healthy individuals and patients are exposed to the new drug under strictly controlled conditions.

Financial Aspects. A budget should cover all aspects of operating the company, including office or laboratory rent, personnel, supplies and materials, any equipment, communications, business development expenses, protection of intellectual property, advertising, and a host of smaller items particular to the start-up situation, such as travel to meet with potential investors. Finally, a forward-looking statement addressing cash flow is expected. In the case of a product that can be immediately marketed, a positive cash flow may be indicated very early in the life of a company. In the event that a device or new medicine needs to be approved, a more sophisticated model might be shown. Such a model would include predicted costs of clinical trials incurred through phases I to III and marketing and sales predictions and costs associated with trials before a positive cash flow is realized; this may not be for 10 or 15 years.

4.4.5
Organizing the Venture

There are various types of company organization; these include the sole proprietorship, the partnership, the limited liability company (LLC), and the corporation, which can be either a C or an S corporation. The advice of an attorney is helpful to determine the best fit for the particular situation. Briefly, if the company will remain relatively modest and not require external funding to start and grow, a "subchapter S corporation" might provide adequate protection, or an LLC construction. If it is anticipated that equity in the company will be sold to raise the capital needed, then a "subchapter C" construction might be a better approach.[1] Other considerations include who controls the organization; how profits, losses, and taxes are handled; whether ownership of the organization can be easily transferred; and how liability will be addressed.

Because entrepreneurs tend to be solo operators who come up with ideas, they often initiate the process of forming a company to exploit the idea, and then move on to new ideas (and to form new companies). However, it is good to assemble a team to develop the idea earlier rather than later because a wide range of skills is needed in forming a growing company. For the science-educated entrepreneur, it would be advantageous to recruit a business-oriented individual early, and vice versa. The early team will probably consist of the science entrepreneur and a business, law, or finance partner. Since legal and financial professionals are available for hire as needed, a partner who has an MBA could be very helpful, especially to perform the market analysis.

The Chief Executive Officer (CEO) in the start-up phase is often the technical person who founded the company. The CEO is responsible for the overall health and direction of the company. Later, when the company matures, it is not uncommon for an experienced business person to assume the leadership of the company as soon as serious funding has been identified; this changeover is often requested by the investors themselves and gives the founder the chance to revert to focusing on the technology. The Vice President for science and/or technology or the Chief Scientific Officer (CSO) is responsible for the science/technology aspects of the company. He or she makes sure that rapid progress is made and that the intellectual property is protected appropriately. A Chief Financial Officer (CFO) is required early in a company's development; this role can be outsourced, but once outside investments have been made, some accountability will be needed, and it will help if the CFO can join the company. Similarly, a company legal expert responsible for protecting and managing intellectual property should be identified. Initially this work is often outsourced.

A board of directors will be established once funding has been obtained, and investors may require board membership as part of their investment strategy. The officers of the company will be employees of the board, which will set goals

---

1) The reference is to Subchapters C and S of Chapter 1 of the US Internal Revenue Code.

for the company. However, a very early-stage organization may not have a board of directors but may have a board of advisors. A scientific board may also be appointed at some stage. Boards are usually kept quite small, but there are no strict rules about this.

Because of the basic nature of the start-up and its usually limited resources, it is important that some consideration be given to what experience additional employees might have with early-stage companies. Frequently, limited cash will also mean that salaries will have to be deferred until either a revenue stream is identified or a financing round is secured. The value of any employee, whether for a small or large company, has to be established and periodically reviewed. This is especially important in the small company because each person represents such a large percentage of the workforce. This is also why flexibility is so important; the ability to wear many hats, i.e., do many and varied jobs, is essential. Small companies do not have the resources to have a specialist for each function. Even CEOs often recall that early on, when resources were limited, they did every job in the company including being the janitor.

4.4.6
Financing the Business

There are a number of ways to raise funds needed for the start-up company, including:

- Self-funding
- Loans
- Grants
- Private investors
- Venture capitalists
- Other types of support

The funding required to start and support the new company is dependent upon the type of business that is being started. If a small amount is required, then the founder may elect to fund it using his or her own resources or a loan. If more resources are needed, an outside source will be required for initial financial support.

Self-Funding. Self-funding a new company means taking responsibility for funding all start-up activities, and it is generally referred to as "bootstrapping" a company. For example, a service company with minimal technology development expenses and an expected early revenue stream could possibly be successfully bootstrapped. Generating an early revenue stream through service work can be a very successful approach to starting and growing a new company. This somewhat traditional approach has the advantage that the founders will be able to keep all of the issued stock and maintain control of the company.

Friends and Family. Using funds from friends and family is also a common approach. Federal regulations address financing from friends and family and limit the number of partners who can be involved. There may be a personal downside to borrowing and risking family member's life savings, and this must be a personal decision.

Loans. Loans from banks and other financial institutions can also be used to fund a company. Usually collateral is required, and banks may favor one type of business organization over another.

Grants. Grants are available to support various stages of research and development. At the federal level, the Small Business Innovation Research (SBIR) grant program provides initial support (phase I) of up to $100 000 and follow-up support (phase II), which can be up to $ 1 million and more. These awards are only available to small for-profit businesses within the United States, and the latest information should be obtained, as these rules change periodically. Institutional grants are also available. The Gates Foundation has provided generous support for research addressing a variety of diseases, including tuberculosis (which infects one-third of the world's population), malaria, and HIV. Finally, local grants may be available to support a new local company.

Private Investors. Angel investors are either individuals or loosely formed groups that have surplus personal funds they are willing to invest in an early-stage company. Angel investors are frequently highly interested in the technology or business plan but are usually not involved in the day-to-day operation of the company. Investors may require part ownership in the company as compensation for risking their resources. This requires assessing the value of the company, the so-called valuation. Valuation is a complicated process and depends on a number of variables, including the value of the intellectual property, the status of the competition, and the potential market for any product.

Venture Capitalists. Venture capitalists pool and manage funds from wealthy individuals and organized investments such as pension funds and use these funds to invest in private companies. The venture capital (VC) group may assist with the expansion of the company to a point where the investors can exit, for example within a five-year period after the initial investment. VC investors will frequently request a formal position on the board of directors through which they can monitor the progress and influence the path chosen for the company.

Other Types of Support. Small businesses have a large effect on the US economy, and there are considerable resources available to entrepreneurs both from the public and private sectors. For example, in North Carolina there are two organizations in particular that are very supportive of early-stage biotechnology-related companies: the Council for Entrepreneurial Development and the North Carolina Biotechnology Center. In addition to these not-for-profit organizations, there are a number of state-funded initiatives, including the North Carolina Small Business and Technology Development Center. Other states have similar organizations.

Local support can take the form of resources from universities and other places of higher education. These resources may include know-how from academics, who may also be interested in collaborating, or facilities such as a nuclear magnetic resonance spectrometer, which may be available for a small fee, thus saving on considerable infrastructure expenses. In addition, university libraries are invaluable. Finally, the availability of trained scientific staff may be an important consideration, and these may be easily found in a university town environment.

4.4.7
Start-Up Dynamics

The ideal scientist for the small company is not necessarily the same scientist who would thrive in a large company environment. The start-up company is in need of entrepreneurial scientists initially. Later, as the company grows, a mix or maybe even a predominance of the traditional scientist's capabilities becomes more important. Some of the possible characteristics for an entrepreneurial scientist and a traditional scientist are listed in Table 4.1.

The characteristics listed indicate trends rather than absolute values. For example, being team-oriented or company-oriented means that the individual's primary goal is the success of the organization, company, or division rather than a particular project, whereas being project-oriented means that the focus is on the success of the individual project. It does not mean that a project-oriented person can not be a team player if needed, nor that an entrepreneurial scientist is unable to operate independently. The entrepreneurial scientist may display a focus on the actual results of a given study rather than being motivated by the process of obtaining those results. He or she may also be broadly interested in a wide range of factors that lead to the result rather than a single specialty. Risk taking is also a characteristic of the entrepreneurial scientist.

During the period that the new business is being established, resources are usually limited. This can create challenges in the retention of good employees. It is important to compensate those individuals well, and this may be done with stock options. Because these are not easily converted into cash when the company is not publicly traded, this is a viable option for the company. As the company becomes successful, an initial public offering (IPO) may occur, and stock options are, therefore, a good incentive for retention.

Early on in the process of starting a company, there will be many different things going on. At least for part of the time the business will be surrounded by chaos. The typical entrepreneur is comfortable with this and recognizes that this is normal. He or she knows that there is direction to this chaos, and the

Table 4.1 Entrepreneurial and traditional scientists.

Entrepreneurial scientist	Traditional scientist
Company-oriented	Project-oriented
Results-motivated	Analysis-oriented
Broadly interested	Specialist
Risk taker	Failure avoider
Motivated by group success	Motivated by contribution to specialty

end can be anticipated as the various goals are completed. It is common for multiple ideas to be simultaneously evaluated and for the direction of the company to change, maybe overnight, with some or all of the projects being altered or dropped. Not everyone thrives in the chaos of an early-stage company. It is important to remember that this is no reflection on an individual as a scientist or a person.

4.5
Timeline

Session 1: The Entrepreneur and the Idea: How to Evaluate It
Session 2: The Opportunity and the Business Plan; Organization
Session 3: Financing the Business, Start-Up Dynamics
Session 4: Student Work
Session 5: Elevator Speech
Session 6: Student Work
Session 7: Student Work
Session 8: Presentation of Student Companies

4.6
Study Plan and Assignments

4.6.1
Session 1

4.6.1.1 **Presentations**
Characteristics of an Entrepreneur (Presentation #12)
From Idea to Concept Evaluation (Presentation #13)

4.6.1.2 **Assignment #1**
Students are asked to think about an idea they may have had for a business or to borrow a business idea in which they are interested. The idea should be in the general area of biotechnology. Students should develop a one-page written description of the basic idea together with the business model, e.g., "We are going to develop a new way to test for drug-resistant bacteria and sell it to hospitals". This does not need to contain a wealth of scientific detail but should give enough information for a potential investor to be able to decide quickly if this is an interesting idea. Each idea will be given a short title, and all ideas submitted to all students. Before the next session, students should rank the ideas from the most interesting to the least interesting. Students should submit their rankings to the instructor. The instructor will collate the results, and at the next session key projects will be identified.

Students should read the Introduction to this chapter.

4.6.2
Session 2

4.6.2.1 **Presentations**
Assessing Opportunity and Writing a Business Plan (Presentation #14)
Organizing the Venture (Presentation #15)

4.6.2.2 **Assignment #2**
The top ideas will be discussed, and projects will be identified. This process should include such aspects as the students' own interests, the novelty and potential of the idea, and should be of an appropriate scope. Students will be divided into groups (maximum of five students per group). Students should assess each idea by first developing a procedure to evaluate the idea, then evaluating the idea. They should also present a team of managers and managers' responsibilities and deliverables.

4.6.3
Session 3

4.6.3.1 **Presentations**
Financing the Business (Presentation #16)
Start-Up Dynamics (Presentation #17)

Student presentations: students should be able to discuss their evaluation of an idea from both a business and a science perspective. Obviously, at any stage a project idea may be determined to be nonviable or inappropriate. Students working on a nonviable project should be reassigned to another project. Approximately 3–5 students should be working on a project.

4.6.3.2 **Assignment #3**
Students should start to write a business plan for their project. This should include a cover page with a name for the business, a list of authors, a list of contents, an executive summary, a short description of the technical idea together with a listing of the competition and a brief outline of how this is going to be a business, i.e., how revenues will be generated. Students will begin to understand that for the project to work best, each student is going to have to assume a particular role (CEO, CSO, CFO, etc.) and be responsible for a defined part of the plan.

4.6.4
Session 4

Students have now organized their companies, electing officers and assigning roles. Students continue work on the business plan.

4.6.4.1 Assignment #4

Students should prepare to present a draft business plan orally to the entire group in week 5. One member of the team now is the spokesperson and will be describing the business plan. The spokesperson can call upon the management team to describe the section of the business plan for which they have responsibility. Students and the instructor will provide feedback and other ideas.

4.6.5
Session 5

4.6.5.1 Presentations

Students present their draft business plans. Students receive feedback and continue to develop the business plans.

4.6.5.2 Assignment #5

A concise oral presentation (slide show) should be developed for the business; this is commonly referred to as the "elevator speech" and should be presented by the CEO in week 6. Each project business will need to identify at least start-up funds and should have some idea of what the first two years will be like regarding growth and goals.

4.6.6
Session 6

4.6.6.1 Presentations

Elevator speeches for each of the projects will be presented and discussed by the students. The formal part of the session includes a discussion of how the business might be funded and grown.

4.6.7
Session 7

4.6.7.1 Assignment #6

Students complete the business plan and prepare a short presentation about the project.

4.6.8
Session 8

Student presentations

Resources

Adams, J. *Investment Biker*. Adams Media Corporation, Holbrook, MA, 1994.

Adams, R. *A Good Hard Kick in the Ass: Basic Training for Entrepreneurs*. Crown Business, New York, 2002.

Bernstein, P. L. *Against the Gods: The Remarkable Story of Risk*. Wiley and Sons, New York, 1996.

Braben, D. W. *Pioneering Research: A Risk Worth Taking*. Wiley and Sons, Hoboken, NJ, 2004.

Cohen, H. *You Can Negotiate Anything*. Bantam Books, New York, 1980.

Dicks, J. W. *How to Incorporate and Start a Business in North Carolina*. Adams Media Corporation, Holbrook, MA, 1997.

Harper, D. *Investing in Biotech*. Raincoast Books, Vancouver, 2002.

Lesonsky, R. E. (ed.) *Start Your Own Business*. Entrepreneur Press, California, 2001.

Malkiel, B. G. *A Random Walk Down Wall Street*. W. W. Norton and Company, New York and London, 1996.

Ormerod, P. *Why Most Things Fail*. Faber and Faber, London, 2005.

Patton, B. M., Ury, W., Fisher, R. *Getting to Yes*. Houghton Mifflin Books, New York, 1991.

Turner, M. L. *The Unofficial Guide to Starting a Small Business*. Macmillan, New York, 1999.

Ury, W. *Getting Past No: Negotiating Your Way from Confrontation to Cooperation*. Bantam Books, New York, 1991.

5
Introduction to US Patent Law [1]

Elaine T. Sale

Contents

1) Disclaimer: The information and opinions provided in this chapter should not be construed as legal advice, nor do they necessarily represent the views of the author's employer. The reader should seek legal assistance where appropriate. The author shall not be liable for any loss of profit or any other commercial damages, including, but not limited to special, incidental, consequential or other damages.

Industry Immersion Learning. Real-Life Industry Case-Studies in Biotechnology and Business
L. Borbye, M. Stocum, A. Woodall, C. Pearce, E. Sale, W. Barrett, L. Clontz, A. Peterson, J. Shaeffer
Copyright © 2009 WILEY-VCH Verlag GmbH & Co. KGaA, Weinheim
ISBN: 978-3-527-32408-8

5.1
Mission

The purpose of this chapter is to provide a general overview of patents and their importance to business and transactions. Although this chapter focuses on United States patent law, students and instructors should be aware that many companies file patent applications in strategically selected countries throughout the world.

5.2
Goals

While it is not expected that students will be able to draft or prosecute a patent application, this chapter should provide a sufficient understanding of patents and the patent process to enable students to recognize new inventions, take initial steps in protecting intellectual property, and understand the basics of intellectual property transactions.

5.3
Predicted Learning Outcomes

Students will

- gain a general understanding of patents and patentable matter;
- learn about the patent process and its importance to business and transactions;
- be able to recognize new inventions;
- know which initial steps to take to protect property;
- understand the basics of intellectual property transactions;
- understand when infringement takes place;
- practice negotiation for a cross-license between two parties.

5.4
Introduction

5.4.1
Brief Description of US Patents

A patent is a grant by the federal government giving an inventor the right to exclude others from making, using, selling, offering to sell, and importing her invention for a certain limited time.[2] In return, the inventor must provide a patent application describing how to make and use the invention. Publication of patent applications by the patent office benefits the public by contributing to the general base of knowledge and possibly leading to further innovation. Accordingly, patents act as an incentive for the inventor to disclose her invention to the public rather than keeping it secret. The length of the exclusionary period, or "patent term", is currently 20 years from the date of filing of a patent application.

Patents and other forms of intellectual property have become increasingly more important to corporations. To illustrate, at the time of this writing, 49 936 US patents had been issued to IBM, 26 146 US patents to Samsung, 3518 US patents to Pfizer, and 1519 US patents to Genentech.[3] The United States Patent and Trademark Office (USPTO) recently estimated that the value of US intellectual property exceeds US$ 5 trillion.[4] This is reflected in the rising number of patent applications filed each year. For example, 147 500 PCT international patent applications were filed in 2006, as compared to 93 237 filings in 2000.[5] The USPTO received over 440 000 patent applications in 2006, and from its inception through 3 June 2008 has granted 7 383 587 utility patents.[6] Most high-technology corporations aggressively pursue patent protection in the US and in other strategically selected countries, with the patent offices of Japan, the US, the Republic of Korea, and China and the European Patent Office (EPO) receiving the greatest numbers of patent filings.[7]

Patents can generate revenue by licensing, sale, or enforcement. While only a small fraction of patents are actually litigated, the amount at stake can be quite large. For example, in a highly publicized dispute, NTP, Inc. filed a patent infringement suit against Blackberry manufacturer Research in Motion, Inc. (RIM). The parties entered into a settlement agreement in 2006, wherein RIM

---

2) Title 35 United States Code Section 271 (35 USC § 271). The complete US Code is available at http://www.law.cornell.edu/uscode (last accessed 15 September 2008).

3) Source: United States Patent and Trademark Office, Patent Full-Text and Full-Page Image Databases (available at http://patft.uspto.gov.).

4) USPTO press release #08-16, 14 April 2008, available at www.uspto.gov/web/offices/com/speaches/08-16.htm.

5) Source: the WIPO Patent Report, 2007 Edition (see http://www.wipo.int/ipstats/en/statistics/patents/patent_report_ 2007.html #P812_44234) (last accessed 15 September 2008). For "PCT", see "The Patent Process: Obtaining and Maintaining a Patent" below.

6) Source: United States Patent and Trademark Office Patent Database (available at www.USPTO.gov).

7) Source: the WIPO Patent Report, 2007 Edition.

agreed to pay NTP US$ 612.5 million to settle all patent infringement claims and for a fully paid-up license.[8]

The possibility of obtaining the right to exclude others from practicing an invention is a strong incentive to invest in research that might otherwise be cost-prohibitive. For example, the Tufts Center for the Study of Drug Development estimated in 2003 that the capitalized cost of developing a new drug was over US$ 800 million (in 2000 dollars), including the costs of failures.[9] A subsequent study by the Bureau of Economics of the Federal Trade Commission estimated the cost at US$ 500 million to US$ 2 billion per drug, depending on the type of therapy.[10]

Without the promise of the exclusive right to make and sell a drug during the patent term, pharmaceutical companies would be reluctant to invest in drug research and development, as it would be less likely that they could recoup their investment without a period of exclusivity from competition from generic drug manufacturers. Upon expiration of a pharmaceutical patent, generic drugs can enter the market at a lower retail price and for a fraction of the cost incurred by the original drug developer. The price of the brand name drug then typically drops in order to remain competitive, and market share is lost to the generic.

The legal basis for patents and copyrights can be found in the US Constitution, which grants Congress the power "to promote the Progress of Science and useful Arts by securing for limited Times to Authors and Inventors, the exclusive Right to their respective Writings and Discoveries."[11] In accordance with this constitutional power, Congress enacted the patent laws, which are codified as United States Code Title 35 (35 USC §§ 1–376). The process of obtaining and maintaining a US patent is further regulated by Title 37 Code of Federal Regulations (37 CFR) – Patents, Trademarks and Copyrights, and is subject to the *Manual of Patent Examining Procedure* (MPEP). Links to 35 USC, 37 CFR, and the MPEP are available online from the USPTO Web site at www.uspto.gov/main/patents.htm.

There are three types of US patents: utility patents, design patents, and plant patents. A utility patent, which protects the functional aspects of products and processes, is the most common type of patent. Design patents protect the ornamental, nonfunctional aspects of articles of manufacture. Plant patents protect new varieties of asexually reproducible plants, including cultivated sports, mutants, hybrids, and newly found seedlings. Tuber-propagated plants and plants found in an uncultivated state are not protected by plant patents (35 USC § 161). This chapter will focus on utility patents.

---

8) See press release dated 3 March 2006 at www.rim.com (last accessed 15 September 2008).

9) DiMasi, J. A., Hansen, R. W., Grabowski, H. G. "The price of innovation: new estimates of drug development costs." *Journal of Health Economics* 2003; 22:151–185.

10) Adams C. P., Brantner V. V. "Estimating the cost of new drug development: is it really 802 million dollars? *Health Affairs (Project Hope)* 2006; 25:420–428.

11) US Constitution, Article 1, Section 8, Clause 8.

In order to obtain a US utility patent, the inventor(s) must file a patent application with the United States Patent and Trademark Office. A patent examiner then reviews the application to determine whether the legal requirements for patentability are met. These requirements are: (1) patentable subject matter; (2) utility; (3) novelty; (4) nonobviousness; (5) enablement; (6) written description, and (7) claims that particularly point out and distinctly claim the subject matter that the applicant regards as her invention.[12]

5.4.2
Patentable Subject Matter

Patentable subject matter is described in 35 USC § 101 as follows: "Whoever invents any new and useful process, machine, composition of matter; manufacture,[13] or any new and useful improvement thereof, may obtain a patent therefore ... ." While "anything under the sun that is made by man" is potentially patentable,[14] certain subject matter has been excluded from patentability by the courts or by statute. The sole statutory exclusion to patentable subject matter is set forth in the Atomic Energy Act, which bars patents on "any invention or discovery which is useful solely in the utilization of special nuclear material or atomic energy in an atomic weapon."[15]

Judicial exclusions of patentable subject matter include mere arrangements of printed matter not functionally related to other claim elements; naturally occurring products that are substantially unaltered; abstract concepts; scientific principles or laws of nature (such as $E = mc^2$); and natural phenomena such as electricity and magnetism. For example, the US Supreme Court has held that substantially unaltered naturally occurring products, such as "a new mineral discovered in the earth, or a new plant found in the wild is not patentable subject matter ... . Such discoveries are manifestations of ... nature, free to all men and reserved exclusively to none."[16]

However, products of nature can be patented if they are purified, isolated or altered.[17] For example, in *Diamond v. Chakrabarty*, the Supreme Court considered whether bacteria genetically modified to contain a plasmid conferring the ability to degrade oil are patentable subject matter. The court concluded that genetically altered bacteria are patentable because they are not products of nature, but rather are made by man.[18] Similarly, *isolated* DNA (e.g., a cloned gene,

---

12) 35 USC §§ 101, 102, 103, and 112.
13) An article of manufacture is anything made from raw materials or other products not thought of as compositions of matter or machines.
14) *Diamond v. Chakrabarty* Volume 447 United States Reports page 303 (447 US 303) (1951) (citing Hearings on HR 3760 before subcommittee No. 3 of the House Committee on the Judiciary, 82nd Cong 1st Sess, 37).

15) 42 USC § 2181 (a).
16) *Diamond v. Chakrabarty*, 447 US 303 (1980).
17) One example of a product of nature that is *not* considered substantially altered is a shrimp with the head and digestive tract removed (i.e., a cleaned shrimp). *Ex parte Grayson*, Volume 51 United States Patents Quarterly page 413 (51 USPQ 413) (Pat Off Bd App 1941).
18) *Diamond v. Chakrabarty*, 447 US 303 (1980).

cDNA, etc.) from any organism is patentable, provided that it meets the other requirements for patentability (i.e., utility, novelty, nonobviousness, written description, enablement, and is properly claimed).

5.4.3
Utility

An invention must be useful, as required by 35 USC § 101. In order to meet the utility requirement, the claimed invention must be operative and have at least one credible use.[19] A credible utility must either be disclosed in the patent application or apparent to one of ordinary skill in the art. For example, a compound with no known use would not meet the utility requirement. Similarly, a claim to a perpetual motion machine would lack utility because it is in conflict with basic scientific principles and therefore not credible. However, as stated by US Court of Appeals for the Eighth Circuit, the utility requirement can be satisfied even if an invention does not work particularly well:

> A small degree of utility is sufficient. The claimed invention must only be capable of performing some beneficial function .... An invention does not lack utility merely because the particular embodiment disclosed in the invention lacks perfection or works crudely .... Nor is it essential that the invention accomplish all its intended functions ... or operate under all conditions ....[20]

For inventions in "predictable" arts, such as the mechanical art, a credible use is typically readily apparent to one of skill in the art and need not be actually demonstrated. In contrast, evidence of utility may be required for inventions in less predictable arts, such as the biotech and pharmaceutical arts. Speculative uses do not necessarily satisfy the requirement for credible utility. In *Brenner v. Manson*,[21] the Supreme Court considered the utility of a claim to a method of making an analog of a steroid known to inhibit the growth of tumors in mice. Because it was not known whether the analog was useful, the court held that the claimed method lacked utility, stating, "A patent is not a hunting license. It is not a reward for the search, but compensation for its successful conclusion."[22] However, actual demonstration of an asserted utility is not always required. For example, it may be possible to obtain a claim to a therapeutic method of using a compound to treat a particular human disease based on *in vitro* or animal data, provided that the *in vitro* system of animal model is considered predictive of the human disease.

---

19) *Manual of Patent Examining Procedure* (MPEP) 2107, which is available at http://www.uspto.gov/web/offices/pac/mpep/mpep.htm (last accessed 15 September 2008).

20) *E. I. du Pont de Nemours & Co. v. Berkley & Co.*, Volume 620 Federal Reports, 2nd page 1247 (620 F2d 1247) (8th Cir 1980).

21) 383 US 519 (1966).

22) *Id.* At 534.

The USPTO's utility examination guidelines require that in addition to being credible, the utility must also be substantial and specific to the claimed subject matter.[23] This requirement excludes "throw-away", "insubstantial", or "non-specific" utilities, such as the use of a transgenic mouse as snake food.[24] The USPTO explains that any mouse could be used as snake food and, unless the transgene makes the transgenic mouse a better snake food, then the utility is not specific to the claimed transgenic mouse. Similarly, stating that a particular expressed sequence tag (EST) from maize can be used as a probe to identify maize polynucleotides is not sufficiently specific to that EST, as any maize EST can be used for the same purpose. Rejections of a claim for lack of utility under 35 USC § 101 may also result in a rejection of the claim under 35 USC § 112, first paragraph, on the basis that the disclosure fails to teach how to use the claimed invention.

5.4.4
Novelty

In order to be patentable, an invention must be novel.[25] A novelty-destroying or "anticipatory" reference must disclose, either explicitly or inherently, every element ("limitation") of the claimed invention. Furthermore, an anticipatory reference must be enabling such that it contains "a substantial representation of the patented improvement in such full, clear, and exact terms as to enable any person skilled in the art or science to which it appertains to make, construct, and practice the invention ... ."[26] The prior art and events that defeat novelty are set forth in 35 USC § 102 (a–g). Section 102(a) lists *acts by others* that defeat novelty if they occur before the patent applicant has made the invention. Specifically, 35 USC §102 (a) provides: "A person shall be entitled to a patent *unless* (a) the invention was known or used by others in this country, or patented or described in a printed publication in this or a foreign country, *before the invention thereof by the applicant for the patent ... .*" Thus, if an invention is publicly known or used in the US or described in a printed publication published anywhere in the world prior to the patent applicant's date of invention, then the patent applicant's invention is not novel.[27] Because section 102(a) governs acts that occur *before* the applicant's date of invention, an applicant may rely on her corroborated notebook records or other documents to establish an earlier date of invention.

Section 102 (b) lists events or acts by anyone, including the inventor, that bar patenting if they occur more than one year before an applicant files a patent ap-

---

23) MPEP 2107.
24) *Id.*
25) 35 USC § 101.
26) However, "it is beyond argument that no utility need be disclosed for a reference to be anticipatory of a claim to an old compound." *In re Schoenwald*, 964 F2d 1122 (Fed Cir 1992).

27) However, an invention that is secretly known rather than publicly known would not bar patentability under section 102 (a). For example, an invention that is held as a trade secret by a first inventor would not bar a patent to that invention by a second inventor who discloses her invention to the public by filing a US patent application.

plication, regardless of whether or not the applicant's date of invention was prior to the listed acts or events. Specifically, section 102(b) provides: "A person shall be entitled to a patent *unless* ... . (b) the invention was patented or described in a printed publication in this or a foreign country or in public use or on sale in this country, *more than one year prior to the date of the application for patent* in the United States ... ." The prior art events defined by section 102(b) are referred to as "statutory bars" to patentability. Accordingly, if a prior art event listed in section 102(b) occurs more than one year prior to filing a patent application, the applicant *cannot* use notebook records to establish an earlier date of invention. However, if the act occurs after the invention by the applicant but less than one year prior to filing a patent application, then the inventor may be granted a patent, provided that she can show that her date of invention preceded the act. This is referred to as a "one-year grace period". In contrast to the US, most foreign countries, including those countries that are parties to the European Patent Convention, do not have a grace period. [28] Accordingly, in order to preserve the ability to obtain patent protection in foreign countries, a priority patent application should be filed prior to publication of the invention. It should be noted that a few countries have grace periods of less than one year. For example, Japan has a six-month grace period, but only for certain activities and very limited types of publications by the inventor. [29]

A "printed publication", as used in 35 USC §102(a) and (b), refers to anything that is printed – by anyone – and is available to the public. Availability to the public is the key; it does not matter whether anyone has actually read the publication. Printed publications include, but are not limited to, journal articles, abstracts, textbooks, a thesis indexed in a library, laid open or published patent applications, [30] papers distributed at public conferences, posters displayed at public conferences for a time sufficient to allow copying or photography, grant proposals that are granted and indexed or otherwise publicly available, catalogs, advertising, and the Internet. Printed publications can be cited as prior art references as of the date they become available to the public. For example, the effective date that a journal article becomes available as a prior art reference is the date of mailing or electronic publication, whichever is first. However, documents shared under an obligation of confidentiality are not available to the public and are therefore not printed publications within the meaning of sections 102(a) or (b).

As stated above, while a publication of the invention by the patent applicant that occurs less than one year before filing does not prevent the patenting of inventions and obvious improvements thereof in the US, most foreign countries do not have a grace period and instead have an "absolute novelty" requirement.

---

28) European Patent Convention, Art 54(2).
29) Japanese Patent Act, Art 29(1).
30) *Bruckelmyer v. Ground Heaters, Inc.*, 78 USPQ2d 1684 (Fed Cir 2006) Holding that unpublished figures that were excised from a published Canadian patent application, constitute a "printed publication" under § 102(b) because they remained in the patent file, which was available for viewing in the Canadian Patent Office in Hull, Quebec.

Accordingly, publication of an invention prior to filing a patent application re-sults in the forfeit of patent rights in most countries.

"Public use in this country" as used in 35 USC § 102 (a–b) refers to anything displayed in public view or used publicly or commercially in the United States. An invention may be considered in public use, even though the invention is not discernible to the public, if the inventor is using the invention commercially or has allowed others to use the invention without restriction. Examples where US courts have found that public use of an invention barred patentability in-clude the unrestricted use of corset steels worn under a garment in public for two years prior to filing a patent application and the hidden locking mechanism of a safe in commercial use at a bank. A method of manufacturing a product, even though the method may not be discernable from the product itself, is in public use if the product is in commerce, and the use is controlled by the in-ventor or her assignee. However, a secret public use by a prior inventor will not bar patentability to a later inventor who promptly files a patent application. There is a limited, judicially created experimental use exception to the public use bar, which allows patenting when the public use is primarily to show that the invention works for its intended purpose. Examples of instances where the experimental use exception has allowed patenting after a public "experimental" use include the outdoor testing of a boat dock and a running track.

"On-sale in this country" as used in section 102 (b) means the physical inven-tion (rather than patent rights to the invention) was sold or offered for sale by anyone within the US A license or assignment of an invention is not consider-ed a sale. The Supreme Court has set forth a two-part test for determining whether an offer for sale has been made.[31] First, there must be a commercial offer for sale within the meaning of contract law. Second, the invention must be "ready for patenting." In order for the invention to qualify as ready for pat-enting, it may have been physically made (an actual reduction to practice), or there may be "drawings or other descriptions of the invention sufficiently de-tailed to enable a person skilled in the art to practice the invention."[32]

A patent is barred by 35 USC § 102 (c) if an inventor has abandoned the in-vention by intentionally dedicating the invention to the public or showing intent not to pursue patent protection. Section 102 (d) bars a patent if the inventor obtains a foreign patent based on a foreign application for the same invention filed in a foreign country more than 12 months before filing the US application. Section 102 (e) treats US patents and patent applications, as well as PCT (inter-national) applications that are published in English and designate the US, as prior art as of the filing date, rather than as of the later date of their publica-tion. Section 102 (f) bars anyone but the true inventor(s) from obtaining a pat-ent. Specifically, 35 USC § 102 (f) provides: A person shall be entitled to a patent unless "(f) he did not himself invent the subject matter sought to be patented."

---

31) *Pfaff v. Wells Electronics*, 525 US 55 (1998).
32) *Id.* at 68.

Section 102 (g) codifies the "first to invent" policy in the US and allows a first inventor who filed a patent application after a second inventor to obtain a patent by showing proof that she was the first to invent and was diligent in reduction to practice, provided that she did not abandon, suppress, or conceal the invention. This is one of many reasons it is important to carefully document conception and reduction to practice. In order to determine who is the first to invent of two inventors, the USPTO holds a complex "interference" proceeding before the Board of Patent Appeals. In contrast, foreign countries grant a patent to the first to file of two applicants.

5.4.5
Nonobviousness

In order to obtain a patent, an invention must be both novel and "nonobvious". Otherwise, a patent could be obtained to methods and products that are only trivial variations of the prior art. Specifically, 35 USC § 103 (a) provides:

> A patent may not be obtained though the invention is not identically disclosed or described as set forth in section 102 of this title, if the differences between the subject matter sought to be patented and the prior art are such that the subject matter as a whole would have been obvious at the time the invention was made to a person having ordinary skill in the art to which said subject matter pertains. Patentability shall not be negatived by the manner in which the invention was made.

In order to assess obviousness, "the scope and content of the prior art are ... determined; differences between the prior art and the claims at issue are ascertained; and the level of ordinary skill in the pertinent art are resolved. Against this background the obviousness or nonobviousness of the subject matter is determined." [33] Thus, an invention is obvious if it could readily be deduced by one skilled in the art based on publicly available information. For example, after the first automobile, a *black* Model T Ford, became available, a subsequent patent application for a *red* Model T Ford would be rejected under 35 USC § 103 (a) for obviousness, as a nonfunctional change in color that satisfied a demand in the marketplace would likely be considered an obvious variation (for example, both red and black bicycles and carriages were in the prior art). However, "a patent composed of several elements is not proved obvious merely by demonstrating that each of its elements was, independently, known in the prior art. Although common sense directs one to look with care at a patent application that claims as innovation the combination of two known devices according to their established functions, it can be important to identify a reason that would have prompted a person of ordinary skill in the relevant field to combine the elements in the way the claimed new invention does. This is so because

---

33) *Graham v. John Deere Co. of Kansas City,* 383
US 1, 17–18 (1966).

inventions in most, if not all, instances rely upon building blocks long since uncovered, and claimed discoveries will almost of necessity be combinations of what, in some sense, is already known." [34] Accordingly, it is impermissible to use hindsight by using the patent application as a guide to combining the prior art in order to reach a finding of obviousness.

In order to prevent the use of hindsight in assessing obviousness, previous case law held that there must be some teaching, suggestion, or motivation to combine two or more prior art references to make the claimed invention, and that there must be an expectation of success in making the combination. Motivation to combine may be found: "1) in the prior art references themselves; 2) in the knowledge of those of ordinary skill in the art that certain references, or disclosures in those reference are of special interest or importance in the field; or 3) from the nature of the problem to be solved, leading inventors to look to references relating to possible solutions to that problem." [35]

Recently, in *KSR International Co. v. Teleflex Inc.*, 127 S Ct 1727 (2007), the US Supreme Court held that the teaching–suggestion–motivation (TSM) test should not be applied rigidly, and that it is no longer the only test that may be used to reach a finding of obviousness, stating: "Rigid preventative rules that deny fact finders recourse to common sense, however, are neither necessary under our case law nor consistent with it." The court provided examples of when obviousness may be found without application of the TSM test. For example, "[w]hen a work is available in one field of endeavor, design incentives and other market forces can prompt variations of it, either in the same field or a different one. If a person of ordinary skill can implement a predictable variation, §103 likely bars patentability. For the same reason, if a technique has been used to improve one device, and a person of ordinary skill in the art would recognize that it would improve similar devices in the same way, using the technique is obvious unless its actual application is beyond his or her skill." [36]

However, "rejections on obviousness grounds cannot be sustained by mere conclusory statements; instead, there must be some articulated reasoning with some rational underpinning to support the legal conclusion of obviousness." [37] "[T]he analysis need not seek out precise teachings direct to the specific subject matter of the challenged claim, for a court can take account of the inferences and creative steps that a person of ordinary skill would employ." [38] Furthermore, "[a] person of ordinary skill is also a person of ordinary creativity, not an automaton." [39] "[A]ny need or problem known in the field of endeavor at the time of invention and addressed by the patent can provide a reason for combining the elements in the manner claimed." [40] Since the "desire to enhance com-

34) *KSR International Co. v. Teleflex Inc.*, 127 S Ct 1727 (2007). *KSR International Co. v. Teleflex Inc.*, 127 S Ct 1727 (2007).
35) *Ruiz v. A.B. Chance Co.*, 57 USPQ2d 1161 (Fed Cir 2000).
36) *KSR International Co. v. Teleflex Inc.*, 127 S Ct 1727 (2007).
37) *Id.*, citing *In re Kahn*, 441 F3d 977, 988 (Fed Cir 2006).
38) *Id.*
39) *Id.*
40) *Id.*

mercial opportunities by improving a product or process is universal", there is an implicit motivation to combine "when the improvement is technology-independent and the combination of references results in a product or process that is more desirable ... because it is stronger, cheaper, cleaner, faster, lighter, smaller, more durable, or more efficient." [41] For example, "[w]hen there is a design need or market pressure to solve a problem and there are a finite number of identified, predictable solutions, a person of ordinary skill has good reason to pursue the known options within his or her technical grasp. If this leads to the anticipated success, it is likely the product not of innovation, but of ordinary skill and common sense." [42]

A rejection of a claim under § 103 for obviousness may be rebutted by (1) showing that the examiner has not made a prima facie case (provided sufficient evidence) to prove an invention is obvious; (2) showing that the cited references teach away from the claimed invention; or (3) providing evidence of secondary considerations supporting nonobviousness, such as commercial success, unexpected results (e.g., unexpected synergy, the failure of others to achieve the results of the invention, a long-felt but unresolved need that the invention satisfies, and copying of the invention by competitors. [43] Arguments that the examiner has not made a prima facie case of obviousness include that the combination does not teach every element of the claimed invention, that the cited art does not qualify as prior art, and/or that one of skill in the art would not have been motivated to make the claimed invention, or have had a reasonable expectation of success in doing so.

5.4.6
Enablement

The patent specification, which is the main portion of the patent application that describes the invention, must meet the requirements of 35 USC § 112. Section 112, first paragraph, codifies the written description, enablement, and best mode requirements, providing:

> The specification shall contain a *written description* of the invention, and the manner and process of making and using it, in such full, clear, concise, and exact terms as to *enable* any person skilled in the art to which it pertains, or with which it is most nearly connected, to make and use the same, and shall set for the *best mode* contemplated by the inventor of carrying out his invention. [Emphasis added.]

In order to satisfy the enablement requirement of 35 USC § 112, first paragraph, the patent application must describe how to make and use the invention so that it may be practiced by one skilled in the art without undue experimentation. Without an enabling disclosure, the inventor has not described the inven-

41) *DyStar Textilfarben GmbH & Co. v. C.H. Patrick Co.*, 80 USPQ2d 1641, 1651 (Fed Cir 2006).

42) *KSR International Co. v. Teleflex Inc.*, 127 S Ct 1727 (2007).

43) *Graham v. John Deere Co.* 383 US 1 (1966).

tion so that it will be available to the public once the patent term expires. However, a patent need not teach, and preferably omits, what is known in the art at the time of filing. The patent application must enable the claimed invention at the time of filing and cannot rely on later improvements in the art for enablement. An applicant may provide evidence of experiments performed after the filing date, typically in the form of a declaration, which demonstrates enablement of the claimed invention. In such cases, the examiner will carefully review the experiments described in the declaration to ensure that they correspond to the teaching and guidance provided by the specification. Furthermore, such a showing also must bear a reasonable correlation to the scope of the claimed invention.[44]

An invention can be enabled even if some experimentation is required, so long as the experimentation is not unduly extensive. The "Wands Factors" are frequently considered in order to determine whether undue experimentation is required. They are: "1) the quantity of experimentation necessary; 2) the amount of direction or guidance presented; 3) the presence or absence of working examples; 4) the nature of the invention; 5) the state of the prior art; 6) the relative skill of those in the art; 7) the level of predictability of the art; and 8) the breadth of the claims."[45] Each of these factors must be considered by the examiner.[46] In general, the mechanical, electrical, and chemical arts are considered more predictable than the biotechnology and therapeutic arts. A conclusion of lack of enablement means that based on the evidence regarding each of the above factors, the specification, at the time the application was filed, would not have taught one skilled in the art how to make and/or use the full scope of the claimed invention without undue experimentation.[47]

5.4.7
Written Description

The written description requirement is frequently confused with the enablement requirement, yet it is distinct. The purpose of the written description requirement is to convey that the inventor was in possession of the claimed subject matter at the time the application was filed. For example, in order to claim a particular nucleic acid, the actual nucleotide sequence is typically disclosed in the patent application. However, in some cases, the written description requirement for a claim to a nucleic acid may be satisfied by a biological deposit of a microorganism containing the cloned nucleic acid with a depository recognized by the Budapest Treaty.[48] The American Type Culture Collection (ATCC) is one such recognized depository. Biological deposits made under the Budapest Treaty

44) MPEP 2164.05.
45) *In re Wands*, 858 F2d 731 (Fed Cir 1988).
46) MPEP 2164.01(a).
47) *In re Wright*, 999 F2d 1557, 1562 (Fed Cir 1993).

48) *Enzo Biochem, Inc. v. Gen-Probe Inc.*, 63 USPQ2d 1609 (Fed Cir 2002).

become available to the public after issuance of the patent application.[49] The patent owner is then notified of any requests for such deposits. Biological deposits can also be used to satisfy the enablement requirement if a biological material is necessary to practice the claimed invention and is not readily available to the public.

5.4.8
Best Mode

The "best mode" is a subjective requirement, wherein the patent application must disclose what the inventor believes, at the time of filing the patent application, to be the best way of making and using the claimed invention. The best mode requirement prevents the inventor from describing only inferior modes in the patent application while keeping secret what the inventor believes to be the best mode at the time of filing the application. Because the best mode is subjective, the USPTO does not examine the patent application for this requirement. Rather, during patent litigation, a court may hold that a patent is unenforceable for failure to disclose the best mode.

5.4.9
Claim Drafting

Claims are the most important part of the patent because they define the invention and the legal (exclusionary) rights of the patentee.[50] Accordingly, claims must be "definite" in order to inform the public what will infringe the patent.[51] Infringement is defined under 35 USC § 271 (a) as follows: "Except as otherwise provided in this title, whoever without authority makes, uses, offers to sell, or sells any patented invention, within the United States or imports into the United States any patented invention during the term of the patent therefor, infringes the patent."[52] Infringement occurs when the apparatus, composition, or method at issue meets every limitation of the claims. Here is an example of a claim: "A composition, comprising: recombinant human erythropoietin."

A claim is the object of a sentence starting with the equivalent of "I claim", "We claim", or "What is claimed is". The claim must begin with a capital letter and end with a period. Periods may not be used elsewhere in the claims, except for abbreviations. Most claims have three distinct components: a preamble, a transition, and elements (limitations). Examples of preambles include: "A method of (doing something)", "A composition", "An apparatus", "A nucleic acid", etc. The transition typically uses words or phrases such as "comprising", "consisting

---

49) However, in 2008, the USPTO proposed a new rule wherein the biological deposit would become available upon publication of the patent application.

50) 35 USC § 112, paragraph 2 provides: "The specification shall conclude with one or more claims particularly pointing out and

distinctly claiming the subject matter which the applicant regards as his invention."

51) 35 USC § 112, paragraph 2.

52) Additional acts of infringement are specified in 35 USC § 271 (b, c, e (2), f and g. Infringement is further discussed below, "Patent Infringement".

of", or "consisting essentially of". The claim elements can include components and connections of an apparatus or machine, steps of a method, or components and/or characteristics of a composition. Consider the following hypothetical claim:

> A composition, comprising: compound A and compound Y.

In this claim, "A composition" is the preamble. "Comprising" is the transition word. "Compound A" and "compound B" are the elements or limitations.

The transition phrase determines whether the claim is "open" or "closed" to additional elements. The term "comprising" makes the claim open to additional elements. More specifically, when the transition "comprising" is used in a claim, an infringing composition, machine, or method must have at least all the elements that follow comprising but may also contain additional elements. For example, consider the following hypothetical claim:

> A composition, *comprising*: A, B, and C.

A, B, and C are the claim elements. A composition that infringes this claim *must* contain each of elements A, B, and C, but may also contain additional elements, such as element D. As a more concrete example, consider the following claim:

> A nutritional supplement, *comprising*: chocolate, calcium, and vitamin D.

A nutritional supplement containing chocolate, calcium, vitamin D, *and vitamin K* would infringe this claim. In contrast, a nutritional supplement containing only chocolate and calcium, or only chocolate, calcium, and vitamin K would *not* infringe this claim, because it is missing an element of the claim (i.e., vitamin D).

> Moving from nutritional supplements to biotechnology, consider the claim:
>
> An isolated cDNA encoding human EPO, *comprising* the nucleic acid of SEQ ID NO:1.

In addition to the nucleic acid of SEQ ID NO: 1, this claim encompasses expression cassettes and vectors containing SEQ ID NO: 1 and may also encompass cells recombinantly engineered to contain this sequence. However, this claim does *not* encompass the human EPO gene in its natural form, i.e., in its normal location in the human chromosome. First, a gene in its normal location is not "isolated". Second, since most eukaryotic genes contain introns, the genomic sequence is typically not the same as the cDNA sequence.

The opposite of the term "comprising", the transition phrase "consisting of" closes the claim to additional components. "Consisting of" claims encompass compositions, methods, or apparatuses that contain all the claim elements, but, traditionally, nothing more. Thus, one might avoid infringement of a "consisting of" claim by adding an element to an otherwise infringing composition. For this reason, "consisting of" claims are less desirable that "comprising" claims and are typically used only when necessary to distinguish the claim over the prior act. Consider the following claim:

A composition, *consisting of:* A, B, and C.

In order to infringe this claim, a composition must contain the elements A, B, and C, and in most cases, nothing more. Examples of noninfringing compositions would be: (1) element A plus element B or (2) elements A, B, C, and D. The first composition does not infringe because it is missing element B. The second composition does not infringe because it contains the additional component D.

However, recent case law has held that the claim term "consisting of" does not exclude additional components or steps that are unrelated to invention. For example, in *Norian Corp. v. Stryker Corp.*, the Federal Circuit held that a claim to a dental repair kit consisting of certain chemical elements was infringed by a kit that contained each chemical element as well as a spatula.[53] Although no other chemicals could be included in an infringing composition, a competitor could not avoid infringement by adding a component unrelated to the invention, such as a spatula. The Federal Circuit explained:

> "Consisting of" is a term of patent convention meaning that the claimed invention only contains what is expressly set forth in the claim. However, while "consisting of" limits the claimed invention, it does not limit aspects unrelated to the invention. It is thus necessary to determine what is limited by the "consisting of" phrase.[54]

In *Conoco Inc. v. Energy and Environmental International L.C.*,[55] the Federal Circuit held that impurities that a person of ordinary skill in the relevant art would ordinarily associate with a component on the "consisting of" list do not exclude the alleged product or process from infringement. In this case, the presence of MIBK (a common impurity in industrial alcohols added to prevent application of a liquor tax) in a water-alcohol mixture did not exclude infringement of a claim having a component "consisting of" a water-alcohol mixture. Thus, impurities normally associated with the component of a claimed invention are implicitly adopted by the ordinary meaning of the components themselves.[56]

As another example, consider a specification that discloses the sequence of an EST (SEQ ID NO: 1) that is useful for detecting expression of a gene, wherein increased expression is correlated with the likelihood of developing heart disease. The full-length gene and cDNA sequences have not been identified. Claim 1 recites:

A polynucleotide, *consisting of:* SEQ ID NO: 1, or the complement thereof.

In this case, the full-length cDNA would not infringe claim 1 because it contains additional sequence beyond SEQ ID NO: 1, which is related to the invention.

---

53) *Norian Corp. v. Stryker Corp.*, 363 F3d 1321, 1331-32 (Fed Cir 2004).
54) *Id.* at 1331.
55) 79 USPQ2d 1801 (Fed Cir 2006).
56) *Id.* at 1809.

The transition phrase "consisting essentially of" allows additional elements that would not materially affect the basic and novel characteristics of the invention. Consider the following claim:

A composition of matter, *consisting essentially of:* A, B, C.

In order to infringe this claim, the composition must contain A, B, and C. If X would not materially change the composition, then A + B + C + X is within the scope of the claim and would infringe. However, if X would materially alter the composition, then A + B + C + X would not infringe this claim.

The claims in the previous examples were directed to compositions. The following are examples of method claims from US patent 7 132 109, entitled "Using heat shock proteins to increase immune response."

1. A method of treating a cancer in a subject in need of said treating comprising the steps of: (a) administering to the subject a composition comprising a component that displays the antigenicity of a cancer cell; and (b) administering to the subject an amount of a purified heat shock protein preparation, wherein the heat shock protein preparation comprises purified (i) unbound heat shock protein, or (ii) heat shock protein bound to a molecule that does not display the immunogenicity of the component.
2. The method *according to claim 1,* wherein the heat shock protein preparation comprises a heat shock protein selected from the group consisting of hsp70, hsp90, gp96, calreticulin, and a combination of any two or more thereof.

Claim 1 above is termed an "independent" claim, because it does not refer to another claim. Claim 2 is a "dependent" claim, because it refers back to another claim (i.e., claim 1). Strict adherence to the claim format governed by 35 USC § 112, paragraphs 2–6 is critical. For example, 35 USC § 112, paragraph 4 provides:

Subject to the following paragraph, a claim in dependent form shall contain a reference to a claim previously set forth and then specify a further limitation of the subject matter claimed. A claim in dependent form shall be construed to incorporate by reference all the limitations of the claim to which it refers.

Now consider the following *paraphrased* versions of claims 1, 2, and 6 of US 5 273 995, which are directed to Pfizer's blockbuster prescription drug, Lipitor:

1. Atorvastatin acid, or atorvastatin lactone, or pharmaceutically acceptable salts thereof.
2. A compound of claim 1, which is atorvastatin acid.
6. The hemicalcium salt of the compound of claim 2.

In *Pfizer Inc. v. Ranbaxy Laboratories Ltd.,* Pfizer asserted the above claim 6 of the '995 patent against Ranbaxy. [57] Claim 2 was construed by the court to ex-

---

57) *Pfizer Inc. v. Ranbaxy Laboratories Ltd.,* 79
USPQ2d 1583 (Fed Cir 2006).

clude salts of atorvastatin acid. [58] Claim 6 is directed to a hemicalcium salt, which is exclusive of claim 2. Because claim 6 depends upon claim 2 but does not further limit claim 2, the Federal Circuit held claim 6 invalid for failure to meet the requirements of 35 USC § 112, paragraph 4. The court recognized:

> The patentee was attempting to claim what might otherwise have been patentable subject matter. Indeed, claim 6 could have been properly drafted either as dependent from claim 1 or as an independent claim – i.e., the hemicalcium salt of atorvastin acid. But, we should not rewrite claims to preserve validity. [59]

A good claim strategy covers as many aspects of the invention and potential infringers as possible. Claims should be drafted not only to a new composition but also to methods of making the composition as well as to methods of using the composition. In addition, it is wise to add dependent claims of increasingly narrower scope. If a patent is litigated, it is possible that broader claims may be held invalid, while narrower claims which still cover the accused infringing product are valid. [60] As an example, consider a patent with the following hypothetical claims:

1. A composition, comprising A and B.
2. The composition of claim 1, comprising 10% A and 90% B.

Imagine that during litigation, a prior art reference disclosing a composition containing 50% A and 50% B is found. This prior art reference would invalidate claim 1, above, for lack of novelty. However, claim 2 might still be valid provided it is nonobvious, for instance because a composition containing A and B in these proportions was previously unknown and the claimed proportions resulted in an unexpected benefit.

5.4.10
Invention and Inventorship

There are two legal components to invention: conception and reduction to practice. Conception has been defined as:

> The complete performance of the mental part of the inventive act. All that remains to be accomplished, in order to perfect the act or the instrument, belongs to the department of construction, not invention. It is therefore

---

58) *Id.* at footnote 6 ("Theoretically, a claimed acid could be liberally construed to include the corresponding salts. ... But here, given the absence of the 'pharmaceutically acceptable salts thereof' language which was used in claim 1, the intrinsic evidence would not have supported such an interpretation of claim 2.")

59) *Id.* at 1590, citing *Nazomi Communications, Inc. v. Arm. Holdings, PLS*, 74 USPQ2d 1458 (Fed Cir 2005) (internal quotes omitted).

60) Conversely, although less typically, a broader claim may be valid while a narrower dependent claim is held invalid (e.g., for lack of written description, etc.).

the formation, in the mind of the inventor, of a definite and permanent idea of the complete and operative invention ... .[61]

An inventor must contribute to the conception of an invention. Conception must be complete enough to enable one skilled in the art to reduce the invention to practice without undue experimentation or the exercise of inventive skill. While the Federal Circuit has stated that conception is the "touchstone of inventorship",[62] determination of inventorship can sometimes be difficult. For example, a mental formulation of a desirable result or a problem to be solved is *not* conception. Instead, it is merely a wished-for result rather than a solution. A professor who tells a student to solve a problem, without more, would not be an inventor of the student's solution to that problem. Furthermore, one who, by following the instructions of the inventor, merely reduces the invention to practice is not an inventor himself. In this regard, authors of publications that describe the invention may not necessarily qualify as inventors, as such publications frequently name those who worked on a project but did not make an inventive contribution. Also, one who merely provides background information is not an inventor.

Inventions made by more than one inventor are addressed in 35 USC § 116, which provides:

> When an invention is made by two or more persons jointly, they shall apply for patent jointly ... . Inventors may apply for a patent jointly even though (1) they did not physically work together or at the same time, (2) each did not make the same type or amount of contribution, or (3) each did not make a contribution to the subject matter of every claim of the patent.
> Whenever through error a person is named in an application for patent as the inventor, or through an error an inventor is not named in an application, and such error arose without any deceptive intention on his part, the Director [of the Patent Office] may permit the application to be amended accordingly ... .

Furthermore, 35 USC § 102(f) provides: "A person shall be entitled to a patent unless ... (f) he did not himself invent the subject matter sought to be patented." Accordingly, identification of the correct inventors is a condition of patentability. A patent can be held invalid if the patent application incorrectly omits an inventor (nonjoinder) or names a noninventor (misjoinder). However, as specified in 35 USC § 116, improper naming of inventors can be corrected while the application is pending or after issue by obtaining a certificate from the USPTO to correct errors in inventorship, provided that the errors occurred without deceptive intent.

---

61) *Mergenthaler v. Scudder*, 11 App DC 264, 276 (1897).   **62)** *Ethicon Inc. v. US Surgical Corp.*, 135 F3d 1456, 1460 (Fed Cir 1998).

The second component of invention is *reduction to practice*, which can be performed by the inventor or someone acting on behalf of the inventor. There are two types of reduction to practice, actual and constructive. "Actual reduction to practice" is the making of a working physical embodiment of the invention. "Constructive reduction to practice" is the filing of a patent application with a full disclosure of the invention. Accordingly, there is no requirement to build and test an embodiment of the invention prior to filing a patent application. However, if the patent office doubts that the invention is enabled or has utility, then the applicant may submit evidence demonstrating that the invention is operative.

The United States is the only country in the world where the first to *invent* is awarded the patent. In the rest of the world, the first to file a patent application is awarded the patent. While it might seem that early filing would be advised in all cases, there are both disadvantages and advantages to filing a patent application early in the process. The advantages are that early filing establishes priority, provides a competitive advantage, and allows for earlier publication, public use, and sale. However, there are also disadvantages, including the fact that enablement and utility can sometimes be difficult to establish early in the process of biotech and pharmaceutical research. In addition, future commercial embodiments may be outside the scope of the early filed patent. Furthermore, the patent term will expire 20 years from the filing date of a US nonprovisional or PCT (international) application.

When two patent applications or one patent application and one issued patent have overlapping patent claims, an administrative process called an "interference" can be initiated in order to determine who was the first to invent. The Board of Patent Appeals and Interferences (USPTO) conducts these proceedings. Their decisions can be appealed to the Federal Circuit. Priority is given to the first to conceive of the invention, followed by continuous and reasonable diligence in reduction to practice (constructive or actual). One who is the first to conceive can be the last to reduce to practice, provided that she can demonstrate diligence in reduction to practice during the critical period — from just before the other party's date of conception to her own reduction to practice. Diligence generally requires showing some activity aimed at reduction during the entire critical period.

The following timelines illustrate relevant dates of conception (C) and reduction to practice (R) by two unrelated inventors, inventor A and inventor B, who filed separate patent applications to the same invention:

Case 1: A is the first to conceive and reduce to practice. Therefore, A has priority over B.

Case 2: A is the first to conceive but the last to reduce to practice. A still has priority, so long as A was diligent in reduction to practice between the time of B's conception and A's reduction to practice.

Case 3: A is the first to conceive and last to reduce to practice, but did not exercise diligence in reduction to practice after B's conception. Therefore, B has priority.

Because conception is a mental act, proof of invention, when necessary, requires corroborating evidence showing that the inventor disclosed to others her "completed thought expressed in such clear terms as to enable those skilled in the art to make the invention." [63] Careful written documentation establishing the date of conception and diligence in reduction to practice is strongly recommended. There are many ways of documenting conception and reduction to practice. They include bound notebooks, reports, invention disclosures, dates of submission of abstracts for publication, and the patent application itself. The importance of good record keeping cannot be stressed enough. An inventor should always strive to have continuity, completeness, and integrity in her records. Continuity refers to a strict chronology in a bound notebook, which preferably includes a table of contents. In addition, there should be no blank pages, which are considered gaps in continuity. Completeness refers to a record of ideas; experiments and their results; supplemental records (e.g., photographs, printouts, etc., which should preferably be affixed to a notebook and dated); and cross-referencing for protocols and data that cannot be affixed to a notebook. In general, an inventor should always assume that her audience is less knowledgeable than they are and include all relevant information. With respect to the integrity of the documentation, inventors should strive to avoid any appearance of record tampering, such as changes, erasures, or whiteouts. Corrections should be made by including new correct information in the notebook and referring back to the incorrect information. Also, no data or pages should be removed from any notebook. To corroborate findings in their notebooks, inventors should have each page of their notebook or other inventive documents witnessed and dated by someone who is a noninventor and is capable of understanding what is disclosed.

5.4.11
The Patent Process: Obtaining and Maintaining a Patent

The steps involved in obtaining a patent include: invention, preparation, and filing of the patent application, patent prosecution, and issue. Once a patent issues, it can be enforced until the expiration of the patent term. In addition, patent maintenance fees must be paid in a timely manner in order to avoid abandonment.

Inventors/applicants can file patent applications in the US and foreign countries as well as international applications under the Patent Cooperation Treaty (PCT), which at the time of this writing had 184 member states (i.e., countries). Although PCT applications are published and optionally examined, they are not granted as patents. Rather, the applicant has from 20 to 34 months (depending on the country) [64] from the earliest priority date to convert the PCT application into a national application in each country where patent protection is desired.

---

63) *Coleman v. Dines*, 224 USPQ 857, 862 (Fed Cir 1985).

64) See http://www.wipo.int/pct/en/texts/pdf/ time_limits.pdf (last accessed on 15 September 2008).

This process is termed "entering national phase". The advantage of filing a PCT application is that it delays the expense and complexities of filing in multiple countries. This delay allows the applicant more time to evaluate the value/ potential of the invention and to consider its patentability in light of the PCT examination report.

A patent application includes (1) a specification that must satisfy the requirements of 35 USC § 112, including the written description, enablement, and best mode requirements, (2) a drawing if necessary to understand the invention, (3) an oath or declaration by the inventor(s) stating that he believes himself to be the original and first inventor, and (4) a filing fee.[65] The fee and oath or declaration may be submitted after the specification and drawings are filed.

The suggested content of the specification is described in MPEP 608.01. The specification should describe the "background of the invention", including relevant prior art known to the applicant and, if applicable, references to problems in the prior art that are solved by the applicant's invention.[66] A brief "summary of the invention" should be directed toward the invention rather than the disclosure as a whole. The summary may point out the advantages of the invention or how it solves problems previously existent in the prior art and may point out in general terms the utility of the invention.[67] The specification should contain a "detailed description" including the preferred embodiment(s) of the invention and may include examples/experiments that were actually performed or prophetic. Work actually performed is typically described in past tense, while prophetic examples must only be disclosed in present tense or otherwise clearly indicate that they have not actually been performed.

The specification can be as short as is necessary to describe the invention adequately and accurately. It is not necessary to provide detailed descriptions of what is conventional and generally widely known in the field of the invention. However, where particularly complicated subject matter is involved or where the elements, compounds, or processes may not be commonly or widely known in the field, the specification should refer to another patent or readily available publication that adequately describes the subject matter.[68]

The claims must meet the requirements of 35 USC § 112, which were previously discussed in this chapter. It is useful to search the literature (including patents and published applications) prior to drafting the claims, so that the claims distinguish the invention over the prior art. However, there is no duty to perform a prior art search. Because the claims are the most critical part of the patent applications, most patent attorneys draft the claims first, and then draft the rest of the specification to more fully describe the claimed invention.

There are two basic types of US patent applications: provisional and nonprovisional. A provisional patent application must include a specification, a drawing if necessary to understand the invention, a filing fee, and it must identify the inventors.[69] However, a provisional application need not include claims, nor an

---

65) 35 USC § 111.
66) MPEP 608.01(c).
67) MPEP 608.01(d).

68) MPEP 608.01(g).
69) 35 USC § 111(b).

oath or declaration by the inventor. Further, in contrast to a nonprovisional application, there is no duty of disclosure, and the filing fee is lower. A provisional patent application is not examined or published and will not issue as a patent. The prime benefit of a provisional application is that it acts as a priority document to establish the date of constructive reduction to practice but does not count toward the 20-year patent term. In order for the provisional application to act as a priority document, the applicant must file a subsequent nonprovisional application or PCT application within one year of the filing date of the provisional application and properly claim priority to the provisional application. This one-year period allows the applicant to evaluate the invention, add new embodiments and data to the specification, and assess the commercial potential of the invention. While a provisional application has fewer formal requirements than a nonprovisional patent application, it still must meet all requirements for patentability to act as a priority document, such as utility, nonobviousness, enablement, and description. In contrast to a provisional application, a nonprovisional patent application is examined and published, and can issue as a patent. Further, nonprovisional patent applications can claim priority to a provisional application, other nonprovisional applications, international (PCT) applications, and foreign applications.

A disadvantage of filing a provisional application is that it will delay examination and issuance of a subsequent nonprovisional application by approximately one year. As an example, a patent application filed as a provisional application in 2000 and as a nonprovisional application one year later in 2001 might issue in 2004[70] and would expire in 2021 (2001+20). In contrast, an invention filed first as a nonprovisional application in 2000 might issue in 2003 and would expire in 2020 (2000+20). For biotech and pharmaceutical inventions, the one-year delay is usually not a problem, as the invention often requires further testing or development before it can be marketed. Accordingly, it would be preferable to have patent protection until 2021, rather than a patent that is granted earlier, before the invention can be sold, and then also expires earlier, in 2020. In contrast to this, inventions in the electrical and software arts are often ready to market soon after they are conceived, and having patent protection at product launch is an important consideration. In addition, it is not uncommon for electrical and software inventions to become obsolete prior to the expiration of the patent term.

A typical timeline for filing provisional, nonprovisional, PCT, and national phase applications is as follows:

Time (T)=0:   File a US provisional application.

---

70) In 2007, the average time between filing and issuance or abandonment of utility, plant, and reissue applications was 31.9 months. See United States Patent and Trademark Office Performance and Accountability Report Fiscal Year 2007 (www.uspto. gov/web/offices/com/annual/2007/2007annualreport.pdf.)

T = 12 months: If international patent protection is desired, file a PCT application and/or directly file foreign applications. If a PCT application is filed, the application can enter national stage in the US (and other foreign countries) at 30 months. Alternatively, both a PCT application and a US nonprovisional application can be filed at this time. If only US patent protection is desired, then only a US nonprovisonal application need be filed. All applications should claim priority to the US provisional application filed at month 0.

T = 18 months: Publication of the US nonprovisional, PCT, and foreign patent applications.

T = 30 months: [71]   National phase entry of the PCT application in designated countries or regional patent offices. Examination of the application will follow.

T = ~2–8 years: [72]   Grant of patents on a country-by-country basis.

Before a patent application for an invention made in the US may be filed in a foreign country, a foreign filing license must first be obtained from the USPTO. [73] The license may be obtained either by filing a petition requesting the license or simply by filing a US patent application. Assuming the USPTO does not find that the publication of the patent application may be detrimental to national security, a foreign filing license will be issued. [74] If an application for an invention made in the US is foreign-filed without first obtaining a license, then the corresponding US patent will not be granted or is invalid if granted unless the failure to procure such license was through error and without deceptive intent, and the patent does not disclose an invention that may be detrimental to national security. [75] In such cases, a foreign filing license may be granted retroactively.

As previously described, the patent application must describe the invention in sufficient detail to enable one of skill in the art to make and use the invention. In addition, nonprovisional patent applications and PCT applications must contain claims that must precisely define the limits of what the invention is, or more precisely, what the patentee may exclude others from practicing. Nonprovisional patent applications and PCT patent applications are published 18 months after their earliest priority date.

During the pendency of the nonprovisional patent application, there is a duty of disclosure, candor, and good faith that rests on every individual involved in

71) This date can vary, depending on the country.

72) The pendency of an application from filing to grant or abandonment can be more or less than 2–8 years and varies widely from country to country, as well as by technology area.

73) 35 USC § 184.

74) 35 USC § 181. If the USPTO believes publication of a patent application may be detrimental to national security, then it is made available to the Atomic Energy Commission, the Department of Defense, and/or other government defense agencies, which may determine that the application shall be held secret.

75) *Id.* § 185.

the preparation and prosecution of a patent application (e.g., inventors, assignees, patent attorneys, and patent agents). This includes the duty to bring information material to patentability to the attention of the USPTO. However, there is no duty to search for relevant prior art. Information is material to patentability when it is not cumulative to information already of record in the application, and (1) it establishes, by itself or in combination with other information, a prima facie case of unpatentability of a claim; or (2) it refutes,[76] or is inconsistent with, a position the applicant takes in: (i) opposing an argument of unpatentability relied on by the Office, or (ii) asserting an argument of patentability, 37 CFR § 1.56. Disclosures of prior art are submitted to the USPTO in the form of an Information Disclosure Statement (IDS). A finding of deceptive intent to withhold material information from the USPTO will render a patent unenforceable for inequitable conduct.

Typically about 22 months or more after a nonprovisional patent application is filed, it will be examined by a patent examiner at the USPTO, who will then correspond in written "Office Actions" with the inventor or assignee or a designated representative such as a patent attorney or patent agent.[77] Often, the first Office Action will be a restriction requirement, which divides the claims into groups of independent and distinct inventions. In addition, for biotech inventions containing claims to protein or nucleic acid sequences, the Action may limit examination to up to ten, or often only one, sequence. In a response to a restriction requirement, the applicant elects the group of claims she wishes to have examined and can traverse the restriction requirement. The applicant's other restricted claims can be examined in a divisional patent application that is filed before the parent application [78] issues.

Following election of claims for prosecution, the patent office will examine the elected claims on their merits, according to the patent laws under 35 USC §§ 101, 102, 103, and 112 as discussed above. Infrequently, the examiner will issue a Notice of Allowance without rejecting the claims. More typically, the examiner will issue an Office Action explaining why certain claims are not allowable. As previously discussed, the reasons for rejection may include nonpatentable subject matter, lack of utility, lack of novelty, obviousness, lack of enablement, lack of written description, or an improper claim format. Applicants can respond to an Office Action by canceling the rejected claims or by arguing that the claims are patentable as filed or as amended. Declarations by inventors or other relevant experts may be used to support such arguments.

---

76) A prima facie case is a case in which there are sufficient facts, if uncontested, to reach a holding or conclusion.

77) In 2007, the average time from filing to the first office action ranged from 17.7 to 34 months, depending on the technology center. See www.uspto.gov/web/offices/com/annual/2007/2007annualreport.pdf (last accessed 24 September 2008).

78) Under MPEP 201.04, the term "parent" is applied to an earlier application of an inventor disclosing a given invention. Such invention may or may not be claimed in the first application. Benefit of the filing date of co-pending parent application may be claimed under 35 USC § 120. The term "parent" is not used to describe a provisional application.

When and if all of the pending claims are patentable, the examiner will issue a Notice of Allowance. The allowance rate of patent applications by the USPTO was approximately 54% in 2006.[79] The applicant then has a nonextendible three months to pay the issue and publication fees. The duty of disclosure continues until the date of issue. Any further applications that claim priority to a "parent" application (e.g., a divisional, continuation, or continuation-in-part application) must be filed while the parent application is still pending (prior to issue or abandonment).

A patent issues on the date of publication of the granted patent and can then be enforced during its term. Once a patent is issued, it is no longer "pending", and an applicant can "mark" a patented product with "patented" or "pat." followed by the patent number. Failure to mark a patented product bars the patentee from recovering damages until notice is given to the infringer.[80]

US patent maintenance fees are due at 3½, 7½ and 11½ years after issuance. The maintenance can be paid up to six months after the due date, although a surcharge will be added. Failure to pay these fees results in an early expiration of the patent term. Should this happen to a patent holder, she can petition for reinstatement of the patent by showing that the delay was unavoidable and paying the maintenance fee. Alternatively, for up to 24 months after the six-month grace period, the applicant can petition to have the patent reinstated by stating that the delay was unintentional and paying the maintenance fee and a surcharge. Under the intervening rights provision of 35 USC § 41(c)(2), another party may continue to use, offer to sell, or sell otherwise infringing things that were made, used, imported, or purchased after the six-month grace period expired and before the reinstatement of the patent. In addition, if another party makes substantial preparations during this period, then a court may deem it equitable to allow the continued manufacture, use, offer for sale, or sale of the patented item or continued practice of the patented process.

Certain types of errors in issued patents can be corrected by a Certificate of Correction, Reissue, or Reexamination. A Certificate of Correction corrects mistakes in the patent made by the patent office or by the applicant, such as typographical errors, failure to incorporate amendments made during prosecution, or incorrect inventorship. The responsibility to ensure a patent's accuracy and arrange for the certificate to correct any inaccuracies falls to the patent holder. When a patent is through error (without any deceptive intent) deemed wholly or partly inoperative or invalid by reason of a defective specification or drawing or by reason of the patentee claiming more or less than she had a right to claim in the patent, the patent may be surrendered, and the USPTO may reissue the patent with a new amended application for the unexpired part of the term of the original patent. However, claims may only be broadened for up to two years after issue. During the patent term, the patent owner or other individuals or en-

79) See Strategic Initiatives presentation from the Biotechnology/Chemical/Pharmaceutical Customer Partnership 5 December 2006 meeting by John Doll, Commissioner for Patent, available at http://www.cabic.com/bcp/120506/ (last accessed 16 September 2008).

80) 35 USC § 287 (a).

tities may request reexamination of the patent when there is a substantial new question of patentability based on the prior art consisting of patents or printed publications.

Once a patent issues, it can be enforced for the term of the patent. As stated above, a patent grants the inventor the right to exclude others from practicing her invention for a certain limited time. From 1790 to 1870 the effective term of a US patent was 14 years from the date it issued. Under the Act of 1870, the patent term was extended to 17 years from the date of issue. Patent applicants were able to extend the term of patent protection by filing a series of continuation, continuation-in-part, or divisional patent applications with claims directed to different aspects of the invention, and which claimed priority to the parent application. By spacing the filing times of these applications, the applicant could obtain a series of related patents with successive issue dates, where the term of each patent was 17 years from issue. In some cases, patentees have achieved patent protection extending 40 years or more from the initial filing date. Also, prior to 29 November 2000, US patent applications and pending claims were not published until they issued as a patent, leaving others to wonder when and if any additional patents relating to the invention would issue and what the patent term might be.

In order to comply with the Trade-Related Aspects of Intellectual Property Rights (TRIPS) Agreement[81] reached by members of the World Trade Organization (WTO), the term of patents granted by WTO member countries is now 20 years from the date a patent application is filed. This aspect of the TRIPS agreement was implemented in the US on 8 January 1995. Accordingly, for US patent applications filed on or after 8 June 1995, the term of utility patents and plant patents is 20 years from the date of filing of the nonprovisional or international (PCT) application.[82] Similarly, the term of a patent issuing from a continuation, continuation-in-part, or divisional application filed on or after 8 June 1995 is 20 years from the filing of the earliest parent nonprovisional or PCT application. The filing date of a provisional application does not start the "20-year clock". For applications that were pending on 7 June 1995 and for unexpired patents still in effect on 8 June 1995, the patent term is the greater of 20 years from filing or 17 years from issue. Under certain circumstances, the term of a utility patent can be extended for certain delays in patent examination and grant caused by the USPTO[83] or for portions of time (up to a maximum of five years) that a drug or biologic was under regulatory review by the FDA.[84]

81) See Annex 1C "Trade-Related Aspects of Intellectual Property Rights (TRIPS)" of the Uruguay Round Agreements, available online at http://www.wto.org/English/docs_e/legal_e/legal_e.htm#TRIPs (last accessed 16 September 2008).
82) 35 USC § 154(a). In addition, the filing date of a US provisional application is not used to calculate the 20-year term. Rather the 20-year date begins from the date of filing the US nonprovisional application or an international PCT application designating the US. However, the term of a design patent, which protects the ornamental design of useful objects, is 14 years from the date of grant (35 USC § 173).
83) 35 USC § 154(b).
84) 35 USC § 156.

5.4.12
Patent Infringement

Patents grant exclusive rights that are territorial in scope. For example, a US patent grants the right to exclude in the US but not in other countries. Acts of infringement are codified in 35 USC § 271 (a), (b), (c), (e)(2), (f), and (g). Direct infringement under section 271 (a) is defined as making, using, offering to sell, or selling any patented invention in the United States or importing into the United States any patented invention during the term of the patent.

Under 35 USC § 271 (b), "[w]hoever actively induces infringement of a patent shall be liable as an infringer." In *DSU Medical Corp. v. JMS Co.*, the Federal Circuit held "[t]o establish liability under section 271 (b), a patent holder must prove that once the defendants knew of the patent, they actively and knowingly aided and abetted another's direct infringement." [85] A finding of inducing infringement also requires that direct infringement occurred. For example, one who knowingly sold a nonpatented article with instructions on how to perform a patented process using that article would actively induce infringement, provided that the patented process was actually performed.

Under 35 USC § 271 (c), "[w]hoever offers to sell or sells within the United States or imports into the United States a component of a patented machine, manufacture, combination or composition or a material or apparatus for use in practicing a patented process, constituting a material part of the invention, knowing the same to be especially made or especially adapted for use in an infringement of such patent, and not a staple article or commodity of commerce suitable for substantial noninfringing use, shall be liable as a contributory infringer." The Supreme Court recently explained:

> One who makes and sells articles which are only adapted to be used in a patented combination will be presumed to intend the natural consequences of his acts; he will be presumed to intend that they shall be used in the combination of the patent. [T]he plaintiff has the burden of showing that the alleged infringer's actions induced infringing acts and that he knew or should have known his actions would induce actual infringements. [86]

It is an act of infringement under 35 USC § 271 (e)(2) to submit an Abbreviated New Drug Application (ANDA) for a drug (or a veterinary biological product) claimed in a patent or the use of which is claimed in a patent, if the purpose is to obtain approval to engage in the commercial manufacture, use, or sale of a drug (or veterinary biological product) before the expiration of such patent. For example, pharmaceutical manufacturers who file a New Drug Application (NDA) for approval of a new drug by the FDA are required to identify, or "list" (in the "Orange Book"), any patent that claims the drug or a method of using

---

85) *DSU Medical Corp. v. JMS Co. Ltd.*, (Fed Cir 2006) 2006 WL 3615056, slip. op. **86)** *Metro-Goldwyn-Mayer Studios, Inc. v. Grokster, Ltd.*, 125 S. Ct. 2764, 2777 (2005).

the drug and for which an action of infringement can reasonably be asserted.[87] In order for a generic manufacturer to market a generic equivalent of a brand name drug, the generic applicant must submit to the FDA an ANDA accompanied by one of the following four certifications: (1) that such patent information has not been filed, (2) that such patent has expired, (3) of the date on which such patent will expire, or (4) that such patent is invalid or will not be infringed by the manufacture, use, or sale of the new drug for which the application is submitted.[88]

It is an act of infringement under 35 USC § 271(f)(1) to supply or cause to be supplied in or from the US all or a substantial portion of the components of a patented invention, in such manner as to actively induce the combination of such components outside the United States in a manner that would infringe if such combination occurred in the US For example, assume a widget patented in the US consists of three unpatented components, A, B, and C. It would be an act of infringement under 35 USC § 271(f)(1) for a US manufacturer to ship unassembled components A, B, and C to a foreign country with instructions for their assembly into the patented widget.

It is an act of infringement under 35 USC 271(f)(2) to supply or cause to be supplied in or from the US "any component of a patented invention that is especially made or especially adapted for use in the invention and not a staple article or commodity of commerce suitable for substantial noninfringing use ... knowing that such component is so made or adapted and intending that such component will be combined outside the United States in a manner that would infringe the patent if such combination occurred in the United States".

It is an act of infringement under 35 USC § 271(g) to import into the US or offer to sell, sell, or use within the US a product that is made by a process patented in the US. A product that is made by a patented process will, for purposes of this title, not be considered to be so made after (1) it is materially changed by subsequent processes; or (2) it becomes a trivial and nonessential component of another product. For example, consider the following claim from a hypothetical US patent: A method for making human fibroblast growth factor (FGF), comprising expressing the gene for human FGF in bacterial cells. Under 35 USC 271 (g), it would be an act of infringement to express the gene for human FGF in bacterial cells in a foreign country and then import the FGF into the United States.

In determining patent infringement, one must first determine the meaning of the claim. Claim meaning is construed by a judge based on plain meaning, definitions in the specification, and the prosecution history. In addition, extrinsic evidence of claim meaning (e.g., tests, dictionaries, expert witnesses, etc.) may be introduced. Then the accused method or composition is compared to the properly construed claim. In "literal" infringement, each element/limitation of the claim must be met by the accused composition, device, or method. For

87) The list of all such patent information is compiled in a collection entitled *Approved Drug Products with Therapeutic* Equivalence Evaluations, known as the "Orange Book."

88) 21 USC § 355 (j)(2)(A)(vii).

example, consider the following claim: "A composition comprising: element A, element B, and element C." A composition comprising elements A and B but not C would not infringe this claim, whereas a composition comprising elements A, B, and C would infringe the claim.

The "doctrine of equivalents" is a judicially created doctrine that allows a finding of infringement even when the accused activity or composition is outside the literal scope of the claims. A common test for infringement under the doctrine of equivalents is whether the accused composition, device, or method accomplishes substantially the same function in substantially the same way to achieve substantially the same result. So under the doctrine of equivalents, if there is a composition comprising elements A, B, and C, a composition comprising elements A, B, and D may infringe if D is an equivalent of C.

As another example, consider the discovery of a cDNA sequence encoding human insulin, which is disclosed as SEQ ID NO: 1. The inventor obtains a claim to a polynucleotide comprising SEQ ID NO: 1. A competitor then "designs around" this claim by using a variant of SEQ ID NO: 1 having a single codon change resulting in a conservative amino acid substitution. Depending on the particular circumstances and the prosecution history of the patent, the competitor's nucleic acid may infringe this claim under the doctrine of equivalents.

There is a time limitation on the recovery of damages. Specifically, "no recovery of damages shall be had for any infringement committed more than six years prior to the filing of the complaint or counterclaim for infringement ... ." [89] However, it should be noted that a patent holder may file suit after the patent expires for acts of infringement that occurred during the patent term.

There are several exceptions to infringement, such as the experimental use exception first put forth by Justice Story, who stated: "It could never have been the intention of the legislature to punish a man, who constructed such a machine merely for philosophical experiments, or for the purpose of ascertaining the sufficiency of the machine to produce its described effects." [90] The experimental use exception is rarely applicable, even with respect to infringement by academic institutions. See, for example, *Madey v. Duke University*, 307 F3d 1351 (Fed Cir 2002), where the Federal Circuit held that the experimental use exception did not apply to use of a patented laser by researchers at Duke University.

Another exception to patent infringement is provided in 35 USC § 271(e)(1):

> (1) It shall not be an act of infringement to make, use, offer to sell, or sell within the United States or import into the United States a patented invention ... solely for uses reasonably related to the development and submission of information under a Federal law which regulates the manufacture, use, or sale of drugs or veterinary biological products.

---

89) 35 USC § 286.
90) *Whittemore v. Cutter*, 29 F Cas 554 (CCD Mass 1813).

The courts have broadly interpreted the exception to infringement under 35 USC § 271(e)(1). In *Merck KGA v. Integra Lifesciences*,[91] the court stated, "the contours of this provision are not exact in every respect, [however] the statutory text makes clear that it provides a wide berth for the use of patented drugs in activities regulated to the federal regulatory process." In this case, the court held that section 271(e)(1) "necessarily includes preclinical studies of patented compounds that are appropriate for submission to the FDA in the regulatory process," and that this section may apply to "(1) experimentation on drugs that are not ultimately the subject of an FDA submission or (2) use of patented compounds in experiments that are not ultimately submitted to the FDA."[92]

Defenses to infringement include claim invalidity, unenforceability of the patent, laches, estoppel, and implied license. In almost every infringement suit, the accused infringer will argue that a claim is invalid and/or the patent is unenforceable. However, a patent is presumed valid, and the accused infringer must demonstrate that the claim is invalid by clear and convincing evidence. A claim may be invalid for failure to meet the requirements of 35 USC § 101 (patentable subject matter and utility), § 102 (novelty), § 103 (nonobviousness), § 112 (enablement, written description, best mode, definiteness, proper claim format). Also, a patent may be held unenforceable for inequitable conduct during patent prosecution or patent misuse (e.g., illegally tying the purchase of a patented item to the purchase of unpatented supplies).

A patent owner who unreasonably delays in filing suit after she knows or should have known of infringement is prevented from recovering damages under the equitable doctrine of laches, provided that the delay prejudices the infringer. Furthermore, under the equitable doctrine of estoppel, a patent owner may not recover damages for infringement if she has represented to the infringer, expressly or implicitly, that she will not enforce the patent.

Under the doctrine of exhaustion or "first sale", once a patented article is sold by the patent owner (or under her authorization) without further restrictions, the purchaser and subsequent buyers are free to use and resell that article without infringing the patent.[93] In addition, the purchaser of unpatented equipment from the owner of a process patent may have an "implied license" to use the article to practice the patented process, if (1) the equipment has no noninfringing uses and (2) the circumstances of the sale "plainly indicate that a grant of a license should be inferred."[94]

Remedies available to the patent owner for infringement include an injunction[95] and compensatory damages for either lost profits or a reasonable royalty, together with interest and costs as fixed by the court. There is an affirmative

91) 74 USPQ2d 1801 (2005), available online at www.law.cornell. edu/supct/pdf/03-1237P.ZO (last accessed 16 September 2008).

92) *Id.* at 1805, 1806.

93) *United States v. General Electric Co.*, 272 US 476 (1926).

94) *Bandag, Inc. v. Al Bolser's Tire Sales*, 750 F2d 903, 998 (Fed Cir 1984).

95) In patent infringement suits, a permanent injunction is a court order that prohibits the infringer from further infringing the patent.

duty of care to avoid infringing upon the rights of known patent holders. The term "willful infringement" refers to intentional infringement of a patent when the infringer had knowledge of the patent and had no reasonable basis for believing that his actions were noninfringing. If willful infringement is found, then the court may award treble [96] damages and attorney fees. [97]

5.4.13
Patent Ownership and Intellectual Property Agreements

Appropriate agreements are a key factor in successfully preserving and exploiting intellectual property (IP). This section will briefly review patent ownership and describe some of the most common types of IP-related agreements. As previously discussed, patents are a form of intellectual property, which like other forms of personal property can be assigned or licensed.

Ownership of a US patent gives the patent owner the *right to exclude* others from making, using, offering for sale, selling, or importing into the United States the invention claimed in the patent. [98] However, patent ownership does *not* guarantee that the owner will have the freedom to make, use, offer for sale, sell, or import the claimed invention because there may be other legal considerations precluding such activities (e.g., existence of another patent owner with a dominant patent, failure to obtain FDA approval of the patented invention, an injunction by a court against making the product of the invention, or a national-security-related issue). [99]

In the US, absent an agreement to the contrary, patents and patent applications are owned by the named inventor(s). [100] Ownership of a patent may be assigned by a written agreement (assignment). In cases where there are multiple owners of a patent, each of the joint owners has an undivided fractional interest in the patent. The Federal Circuit has described the rights of co-owners as follows:

> Each co-owner of a United States patent is ordinarily free to make, use, offer to sell, or sell the patented invention without the regard to the wishes of any other co-owner. 35 USC § 262. Each co-owner's rights carry with them the right to license the invention to others, which does not require the consent of any other co-owner ... Thus, unless the co-owner has given up these rights through an 'agreement to the contrary', the co-owner may not be prohibited from exploiting its rights in the patent, including the right to grant licenses to third parties on whatever conditions the co-owner chooses. [101]

---

96) The judge may triple the damage award as a penalty for willful infringement.
97) 35 USC §§ 284–285.
98) 35 USC § 154(a)(1).
99) MPEP 301.

100) *Beech Aircraft Corp. v. EDO Corp.*, 990 F2d 1237 (Fed Cir 1993).
101) *Schering Corp. v. Roussel-UCLAF SA*, 104 F3d 341 (Fed Cir 1997).

While joint patent owners may separately license or assign their patent rights, unless there is an agreement to the contrary, all joint owners are necessary parties to a suit for patent infringement. However, unless a joint patent owner has a contractual obligation to join an infringement suit, she cannot be forced to join the suit.[102] Therefore, in a case where there are two joint patent owners and no contractual agreement regarding patent rights, each may separately license the patent rights without sharing licensing fees and royalties with the other party. Thus, a party seeking a nonexclusive license could negotiate with each owner and accept a license with the best terms from a single owner.

In order to ensure that a company, rather than an employee, is the owner of an invention made within the scope of employment, the company (or research institution) should require as a condition of initial hiring, an employment agreement wherein the employee assigns the rights in all future inventions to the company. In addition, the agreement should require the employee to promptly report inventions to the company in writing and impose a duty not to disclose confidential company information to third parties. Similarly, when two parties collaborate in research, it is wise to first enter into an agreement specifying the ownership and rights in any joint inventions resulting from the research.

A patent assignment is a transfer of (1) all rights in the patent or patent application; (2) an undivided share of patent rights (e.g., all of the interest of one joint inventor); or (3) all rights within a specified portion of the US.[103] The Supreme Court has explained:

> A transfer of either of these three kinds of interests … vests in the assignee a title in so much of the patent itself, with a right to sue infringers; in the second case, joint with the assignor; in the first and third cases, in the name of the assignee alone. Any assignment or transfer, short of one of these, is a mere license, giving the licensee no title in the patent, and no right to sue at law in his own name for infringement. [W]hen the transfer amounts to a license only, the title remains in the owner of the patent; and suit must be brought in his name, and never in the name of the licensee alone … .[104]

As compared to assignment of patent rights, the licensing of a patent transfers a bundle of rights that is less than the entire ownership interest, e.g., rights that may be limited as to time, geographical area, or field of use. As an example of a field-of-use limitation, an owner of a patent to an antibody might license the antibody to one licensee for use in the field of diagnostics and to another licensee in the field of therapeutics. A patent license is, in effect, a contractual agreement that the patent owner will not sue the licensee for patent infringement if the licensee makes, uses, offers for sale, sells, or imports the claimed invention, as long as the licensee fulfills its obligations under the license agreement.

102) *Ethicon, Inc. v. Unites States Surgical Corp.*, 45 USPQ2d 1545 (Fed Cir 1998).

103) *Waterman v. Mackenzie*, 138 US 252 (1981).

104) *Id.* at 255.

Licenses can generate large revenues even before the first product is sold. For example, Enanta Pharmaceuticals entered into a commercialization and development agreement with Abbott Laboratories for two *preclinical* hepatitis C virus (HCV) protease inhibitors to treat HCV infection. Enanta Pharmaceuticals will receive $57 million upfront and an equity investment of up to about $13 million and up to $250 million in milestone payments for the first product.[105] Promising drugs in later-stage clinical trials can generate even greater revenues. In 2005, AtheroGenics and AstraZeneca announced a license and commercialization agreement for AGI-1067, an oral antioxidant and anti-inflammatory compound in phase III clinical trials to treat atherosclerosis. Under the agreement, AtheroGenics will receive an upfront fee of $50 million and will be eligible for up to $1 billion subject to the achievement of specific milestones for development, regulatory, and sales milestones, in addition to royalties.[106]

Assignments or other conveyances of patent ownership or rights should be recorded with the USPTO. This is analogous to recording a deed for real property. An assignment, grant, or conveyance is void against any subsequent purchaser or mortgagee for valuable consideration, without notice, unless it is recorded in the Patent and Trademark Office within three months from its date or prior to the date of such subsequent purchase or mortgage.[107] Recorded assignments of patents and published patent applications can be searched at the following USPTO Web site: http://assignments.uspto.gov/assignments/q?db=pat.

Confidentiality or Nondisclosure Agreements (CDAs) restrict the disclosure of confidential information by the receiving party to a third party. It is critical to ensure that a confidentiality agreement is in place before disclosing confidential information to another party. Confidentiality agreements may be limited in the scope and/or type of information that will be disclosed, the purpose for which the confidential information may be used, the length of time the receiving party must keep the information confidential and the term of the agreement itself. Accordingly, it is important to verify before disclosure that the confidential information is within the scope of the agreement, and that the agreement is still in effect at the time of disclosure. Also, many confidential disclosure agreements require that orally disclosed confidential information must be disclosed in writing to the recipient and marked as confidential within a certain time (e.g., 30 days, 60 days, etc.) of the oral disclosure.

A Material Transfer Agreement (MTA) limits what a recipient may do with proprietary material (e.g., a chemical compound, microorganism, antibody, plasmid, etc.) that is transferred under the MTA. For example, an MTA can restrict transfer of the material or derivatives thereof to a third party, limit the uses of the materials, limit the disclosure of data generated using the material, etc. An

---

105) See press release of 12 December 2006 at http://www.enanta.com (last accessed 16 September 2008).

106) See press release of 22 December 2005 at http://www.atherogenics.com/investor/index.html (last accessed 16 September 2008).

107) 35 USC § 261.

MTA may also grant rights to the owner of the proprietary material, such as an option to license inventions made using the material.

5.5
Timeline

Session 1: Introduction to US Patents, Patentable Subject Matter
Session 2: Novelty, Nonobviousness
Session 3: Enablement, Written Description, and Best Mode
Session 4: Claims
Session 5: Invention and Inventorship
Session 6: The Patent Process: Obtaining and Maintaining Patents
Session 7: Infringement
Session 8: IP-related agreements

5.6
Study Plan and Assignments

5.6.1
Session 1

5.6.1.1 Assignment #1: Preparation for Session 1
Read the Brief Description of US Patents, Patentable Subject Matter and Utility sections from this chapter and *Diamond v. Chakrabarty*, 447 US 303 (1980) (available at http://caselaw.lp.findlaw.com/scripts/getcase.pl?court=us&vol=447& invol=303). This US Supreme Court opinion discusses patentable subject matter.

5.6.2
Session 2

5.6.2.1 Assignment #2: Preparation for Session 2
Read the Novelty and Nonobviousness sections of this chapter and be prepared to discuss *Impax Laboratories, Inc. v. Aventis Pharmaceuticals, Inc.*, 81 USPQ2d 1001 (Fed Cir 2006) (available at http://www.ll.georgetown.edu/federal/judicial/ fed/opinions/05opinions/05-1313.pdf). This opinion discusses the requirements of an anticipatory reference.

5.6.3
Session 3

5.6.3.1 Assignment #3: Preparation for Session 3
Read the Enablement, Written Description, and Best Mode sections from this chapter and be prepared to discuss *In re Wallach*, Slip 03-1327 (Fed Cir 2004),

decided August 11, 2004 (available at http://www.ll.georgetown.edu/federal/judi-cial/fed/opinions/03opinions/03-1327.html). This opinion discusses the written description for nucleic acid molecules.

5.6.4
Session 4

5.6.4.1 Assignment #4: Preparation for Session 4
Read the Claim Drafting section from this chapter and Boczkowski, D., Nair, S.K., Snyder, D., Gilboa E. "Dendritic cells pulsed with RNA are potent antigen-presenting cells in vitro and in vivo." *Journal of Experimental Medicine* 1996; 184(2): 465–472 (available online at http://www.jem.org). Be prepared to discuss possible inventions reported in this article and strategies for claiming those inventions. For the purpose of this assignment, it is not necessary to search the prior art.

5.6.5
Session 5

5.6.5.1 Assignment #5: Preparation for Session 5
Read the Invention and Inventorship section from this chapter.

5.6.6
Session 6

5.6.6.1 Assignment #6: Preparation for Session 6
Read The Patent Process: Obtaining and Maintaining a Patent section from this chapter. Also, navigating the USPTO Web site (http://www.uspto.gov), search for US patent 4 703 008 ("DNA sequences encoding erythropoietin") and answer the following questions:

  Who are the inventors?
  What date was the patent application filed?
  What is the filing date of the earliest priority application?
  What date did the patent issue?
  On what date would the patent term be expected to expire?
  Who was the assignee of the patent when it issued?

5.6.7
Session 7

5.6.7.1 Assignment #7: Preparation for Session 7
Read the Patent Infringement section of this chapter and US patent 4 677 195. Compare the claims of the '195 patent to the claims of the '008 patent assigned last week. How might the claims overlap?

5.6.8
Session 8

5.6.8.1 **Assignment #8: Preparation for Session 8**
Read the Patent Ownership and Intellectual Property Agreements section of this chapter. Consider again the '195 and '008 patents. The owners of these patents are two biotech companies that wish to market competing human erythropoietin products. Assume both companies wish to market recombinant human erythropoietin and that the proposed products fall within the scope of each patent. Students should prepare to negotiate a term sheet for a cross-license of these two patents, one playing the role of Kirin-Amgen, and the other playing the role of Genetics Institute.

Resources

Brunsvold, B.G., O'Reilley, D. *Drafting Patent License Agreements*, 4th ed. BNA Books, Washington, DC, 1998.

Chisum, D., Jacobs, M. *Understanding Intellectual Property Law* (Legal Text Series). Matthew Bender & Co., Inc., New York, NY, 2004.

Schechter, R., Thomas J.R. *Principles of Patent Law* (Hornbook Series). West Publishing Group, St. Paul, MN, 2004.

Thomas, J.R. *Pharmaceutical Patent Law*, 74. BNA Books, Washington, DC, 2005.

6
Intellectual Property Management

William Barrett[1]

Contents

1) Portions of this chapter are adapted from   *Profiting from Ideas in an Age of Global Inno-*
   Barrett, W., Price, C., Hunt, T. *iProperty:*   *vation.* Wiley & Sons, 2008.

Industry Immersion Learning. Real-Life Industry Case-Studies in Biotechnology and Business
L. Borbye, M. Stocum, A. Woodall, C. Pearce, E. Sale, W. Barrett, L. Clontz, A. Peterson, J. Shaeffer
Copyright © 2009 WILEY-VCH Verlag GmbH & Co. KGaA, Weinheim
ISBN: 978-3-527-32408-8

6.1
Mission

This case focuses on the strategic management of intellectual property. The case is based on a hypothetical company with a new technology. Students are expected to develop a strategy and plan for developing an intellectual property portfolio to protect the new technology and for avoiding infringement of others' patents.

6.2
Goals

Students will

- analyze a new technology from an intellectual property perspective;
- develop a patent strategy, including vision, mission, and goals;
- understand the basic economic incentives of intellectual property laws;
- learn about the impact of globalization on the management of intellectual property;

- identify strategically important risks and opportunities arising out of the intellectual property arena;
- develop a patent map for a new technology area;
- practice good business presentation and report preparation skills.

6.3
Predicted Learning Outcomes

Students will learn how to

- evaluate an invention;
- create a patent-claiming strategy;
- search a patent database;
- identify infringement risks;
- map a set of patents;
- present intellectual property information.

6.4
Introduction

Companies compete and grow based on the development of new ideas. These ideas originate in the minds of company employees. How is a company to tap into those ideas, decide which ones to protect, where to protect them, and how to protect them? Suppose Julie, an employee of company A, invents a new idea for a fluorescing protein? How does the mere idea grow into a valuable intellectual property (IP) asset for the company? Moreover, how is the company to know whether Julie's idea is likely to infringe the IP assets of another company?

A successful IP management effort requires forethought and strategy. An IP strategy must be developed in light of an understanding of the legal and economic attributes of IP, the technology being protected, and the competitive environment in which the product will be brought to market. This chapter introduces the core concepts necessary to initiate an IP management program. These concepts include the economics of IP and the globalization of innovation, crucial considerations for any IP strategy. Next, the chapter discusses the development and implementation of an IP strategy. What does it look like? How is it developed? Finally, the principles involved in the development and implementation of an IP strategy, including principles for assessing inventions and developing an understanding of the patents of competitors, are presented. This chapter assumes a basic understanding of IP law, which is presented in Chapter 5 ("Introduction to US Patent Law).

6.4.1
Economics of IP

Any manager of IP should have a working understanding of the basic economics underlying intellectual property policies. It is perhaps easiest to understand the need for IP by considering problems that would occur in the absence of IP protections. Consider Jane, a seller of farm goods, who wants to create an improved plow. She stays up late at night for months studying the physics of plows and working on the new design. She orders expensive new materials and coatings to try on the plow as well as expensive forging equipment and supplies for making experimental plows. She rents land and pays a helper to test the plow designs. She finally completes the new design and offers it for sale in her farm goods store, and her plow sales double as a result. In the absence of any protection for her innovative plow, the owner of the competitive farm goods store next door buys one of her new plows, studies the improvements and immediately starts making and selling an identical plow. In essence, he reaps the benefits of her investment in time and resources. As a result, Jane's sales slump, and Jane never recoups the investment she made in her new plow, while her competitor who has made no such investment, makes a nice profit by copying Jane's ideas.

IP laws are a policy response to this type of market failure. In a purely free market system, our plow innovator would have a disincentive to invest time and resources in innovating due to the likelihood that she would not be able to recoup her investment. Because of this risk, she and others like her might forego the investment in developing new innovations. The new plow and many other innovative products would not be made. Moreover, the benefits to society from the improved plow – improved crop production, less expensive food – might also be lost or at least delayed. While the complexities of the relationship between IP and innovation are still poorly understood, societies have nevertheless designed their IP laws with the goal of motivating investment in innovation.[2] Many economists consider the rewards reaped by innovators to be a kind of "tax," which compensates the innovator for the risk undertaken in developing the new innovation. The tax is paid by the consumers of the patented product in the form of the higher price obtained by the patent owner.

Another important economic attribute of IP, especially patents, is that IP facilitates the sale of ideas. Ideas are notoriously difficult to define. But the patent system has evolved to provide as much precision in definition as human languages will permit. Ideas can thus be defined in patents, and the patents can be sold or licensed – a proposition that is much easier than selling unprotected, undefined ideas. Thus, as illustrated in Figure 6.1, patents facilitate transfer of rights to inventions from one company to another. One advantage of this characteristic of patents is that rights to inventions can be transferred at each stage of the development process. Remember Julie? Her company can patent her in-

---

2) For a contrary view, see Bessen, J. Meurer, M. *Patent Failure: How Judges, Bureaucrats,* *and Lawyers Put Innovators at Risk.* Princeton University Press, Princeton, NJ, 2008.

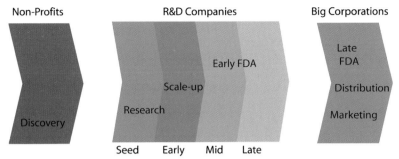

Non-Profits   **R&D Companies**   **Big Corporations**

Late
FDA

Early FDA

Distribution

Scale-up

Research   Marketing

Discovery

Seed   Early   Mid   Late

Fig. 6.1 Patents facilitate transfer of rights to inventions from one company to another ("handoff"). The transferability of patent rights permits flow of rights from nonprofit entities, such as universities, to small re-search-and-development (R&D) companies to big corporations. The nonprofits create the new ideas, the R&D companies take on the initial burden of development, and the big corporations complete the development process and take the products to market.

vention and sell it to another company to bring in revenue or outlicense her patent to another company in exchange for a percentage of their product sales.

6.4.2
Globalization of Innovation

Economic globalization makes it increasingly difficult for companies to compete on the basis of price, especially companies in which the cost of labor is high. Companies can outsource the labor to countries in which labor costs are low. Lowering labor costs provides a temporary price advantage, which is quickly eliminated when other companies do the same. As companies compete to lower price, profits are eventually reduced to a bare minimum or even eliminated altogether. In the latter case, the company must terminate the business, or it will go out of business.

Companies that wish to earn a greater than average return on their investment for a sustained period of time must have a competitive advantage that cannot readily be copied by others. One way to do this is by creating a new technology and using IP to block competitors from accessing the technology. In the global era, companies are increasingly making use of this strategy. Baruch Lev, Professor of Accounting and Finance at New York University, has stated that "given the decreasing economies of scale (efficiency gains) from production coupled with ever increasing competitive pressures, innovation has become a matter of corporate survival." [3] If innovation is a matter of corporate survival, protecting that innovation in order to extend the competitive advantage conferred by that innovation is a critical task for the innovating company.

---

3) Lev, B. *Intangibles: Management, Measurement, and Reporting*. Brookings Institution Press, Washington, DC, 2001, pp. 14–16.

At the same time, globalization has caused an expansion and redistribution of sophisticated innovation capabilities and markets around the globe. These changes offer many opportunities to companies looking to reduce the cost of innovation, but they also complicate the management of IP. Complications result from the fact that (1) companies may move their own centers of innovation to less expensive innovation "hot spots," but at the same time (2) new competitors are emerging in these innovation hot spots. For example, new competitive threats may arise from:

- an outsourcing or off-shoring strategy that locates critical technology in a country where the trade secret or patent laws are poorly defined or poorly enforced;
- a competitor's move to set up manufacturing operations to make a copycat product in an emerging innovation hot spot where there is no adequate patent protection;
- a competitor's move to serve markets where the innovator has passed up the opportunity to obtain patents or where patents are poorly enforced or not enforced at all;
- escape of valuable know-how from an off-shore operation when employees leave to start their own company or join a competitor;
- competition from novel, patent-protected products produced by innovators located in one of the world's many new innovation hot spots.

These and many other potential threats are, if not unique to the global marketplace, certainly increasing in probability due to economic globalization and the increasing global spread of innovative capabilities.

Moreover, companies must work with laws and cultures in countries around the world that impact their ability to protect innovation. Challenges facing global innovators include *external risks*, such as inadequate laws and enforcement; *internal risks*, such as insufficient IP management knowledge or an employee culture that does not appreciate the value of IP; and *technology risks*, such as risk of electronic transmission or Internet publication of IP information. These external, internal, and technology risks may decrease efforts to protect IP appropriately.

6.4.3
Creating an IP Strategy

David A. Aaker, Vice Chairman of Prophet Brand Strategy and Professor Emeritus of the Haas School of Business at the University of California at Berkeley, emphasizes that the goal of any strategy process is to "precipitate as well as make strategic decisions." According to Aaker, "the identification of a strategic response is frequently a critical step. Many strategic blunders occur because a strategic decision process was never activated, not because a wrong decision was made."[4] The same is true for the management of IP. Serious blunders

---

4) Aaker, D. *Developing Business Strategies*, 5th ed. John Wiley & Sons, 1998.

usually occur if the issue was never identified and surfaced for a decision, not necessarily because the wrong decision was made.

Consider bacteriologist Alexander Fleming, who discovered penicillin while working at St. Mary's Hospital in London in 1928. St. Mary's failed to patent the discovery. Andrew Moyer, working for the US Department of Agriculture, later patented the process for making penicillin, permitting US companies to reap the benefit of mass production. Either St. Mary's had no intention to or strategy for raising the patenting question, or it made a strategic blunder. The world got penicillin anyway, but St. Mary's received no financial credit, which it could have invested back into its charitable mission.

Companies wishing to protect their innovations and avoid strategic blunders are faced with a plethora of decisions to make in the IP arena: What to protect? How to identify valuable innovations worth protecting? What forms of protection make sense for each particular innovation? How to avoid wasting resources on unnecessary protection? and so on. The IP opportunities available to innovating companies come in myriad shapes and sizes: utility patents, design patents, trade secrets, and copyrights, as well as practical and contractual protections (see Chapter 5 for a discussion of the forms of IP protection). Moreover, the increasing opportunities for protection mean that more competitors are filing their own patents, creating a patent landscape that is densely populated and potentially complicated and risky. Companies that infringe the IP of others may be subject to court orders that close factories and the payment of large penalties to the owners of the infringed IP.

In response to this complexity, companies must create IP strategies that are designed to ensure the best possible protection of their innovations. The IP strategy, which is expressed in an IP plan, will typically include vision and mission statements. It will also include specific IP objectives that define how the company will invest its time and resources to develop an IP portfolio that will, along with the business and product strategies, help to maintain a sustainable competitive advantage.

6.4.3.1 IP Vision

The most fundamental statement in the IP plan is the IP vision. The IP vision guides all decisions about IP. No decision should be made that fundamentally conflicts with the IP vision.

In their study of visionary companies, James Collins and Jerry Porras identify three components of a business vision: core values, a core purpose, and one or more BHAGs ("Big Hairy Audacious Goals").[5] The same is true for an IP vision. It describes the reason why the company is investing in an IP portfolio and the expected benefit of the portfolio, and includes one or more visionary BHAGs. The IP vision should address opportunities, account for internal cap-

---

5) Collins, J., Porras, J. "Building Your Company's Vision." *Harvard Business Review* September–October 1996.

abilities, inspire commitment, and be aligned with the company's vision. For example, "We will build the premier portfolio of patents in our space/area of expertise. Our patent portfolio will be highly visible and will attract the attention of our competitors as well as our target investors/purchasers/partners. Our sophisticated IP program will establish a zone of protection around our space/area of expertise, will minimize the risk of infringing others' patents, and will dramatically increase our company's valuation."

6.4.3.2 IP Plan

The IP plan outlines the guiding principles and specific strategies and tactics that must be realized to achieve the vision. Examples of issues to consider for an IP plan include:

Portfolio Breadth. Will the IP investment be focused on a core set of IP that directly supports very specific business objectives or will the IP investment be used to build a portfolio that broadly protects multiple product spaces and/or technology spaces?

Information Access. How will the company make crucial information about its IP portfolio and landscape available to business leaders? How will the company facilitate accurate communication of IP information to partners and customers?

Infringement Risk. Will the company try to avoid infringement risks at all costs? Is this even possible? What level of infringement risk is acceptable?

Rights Preservation. How will the company avoid unnecessary loss of IP, e.g., due to public disclosures, offers for sale, and the like?

Corporate Culture. Does corporate management understand and respect the role and value of IP? If not, what specific steps will be taken to influence positive changes in the culture? Is the current level of inventor participation sufficient? Are inventions coming only from a small group of people or are they broadly distributed across the employee base? How will the company stimulate improved participation?

6.4.3.3 IP Team

Because decisions about IP involve a large number of complex factors, the results of an IP management program can be difficult to predict. Among the factors that must be considered are:

1. Technical questions:
   (a) How novel is the technology?
   (b) Is the technology useful in a narrow or broad range of applications?
   (c) How long it will take to develop the technology?

2. Business questions:
   (a) Is there demand in the market for the technology?
   (b) What are the potential sales?
   (c) How much will it cost to develop the technology?
   (d) How long is the product's lifecycle expected to be?
   (e) In what countries would it be manufactured and sold?
3. Legal questions:
   (a) What forms of IP are available?
   (b) Is the technology patentable?
   (c) In what countries is it patentable?
   (d) What countries would be likely to enforce a patent?

Like all decisions, strategic IP planning is done in the context of a mental model. A mental model is an intellectual facsimile of the world in which mental scenarios are played out to ascertain the potential consequences of a variety of potential choices.[6] Decision makers use their rational abilities along with their imaginations to assess what the future might hold for any of a number of potential scenarios, and decisions are made in light of this assessment. Mental models can be complex or simple. Complex mental models account for many variables in playing out the results of various potential decisions. Simple mental models take into account only a few variables. Decisions about IP necessarily require complex mental models because they involve predicting the future based on an interrelated set of legal, technical, and business issues. Consequently, these decisions are best made in the context of an IP team in which the members have a diverse set of expertise covering all the required legal, technical, and business issues. The mental model assembles the collective knowledge of a diverse team about all the technical, business, and legal questions described above.

6.4.4
Invention Assessment

Once an IP plan is in place, each invention must be assessed. It is necessary to have a set of invention assessment parameters to help separate important inventions from unimportant inventions. Examples of such parameters include novelty, probability of technical success, and invention type.

6.4.4.1 Novelty
The novelty parameter relates to how large of an advance the invention has over the state of the art. The purpose of this parameter is to divide a set of novel inventions into groups based on whether they are highly novel, moderately novel,

6) Craik, K. *The Nature of Explanation*. Cambridge University Press, Cambridge, UK, 1943.

Table 6.1 Evaluating novelty.

Rank	Score	Explanation
High	2	The invention is an unusually large advance over the present state of the art
Medium	1	The invention is a solid innovation, a moderate advance over the present state of the art
Low	0	The invention is an incremental advance over the present state of the art

Table 6.2 Evaluating technology risk.

Rank	Score	Explanation
High	2	Overcoming the technical hurdles will be a matter of routine development
Medium	1	There are significant hurdles, but they can be overcome with moderate effort
Low	0	The hurdles are quite large, but with a significant investment of time, resources, and brainpower, they can be overcome

or incrementally novel. Table 6.1 shows an example of how inventions can be ranked.

6.4.4.2 Probability of Technological Success (Technology Risk)

Technology risk separates inventions based on likelihood that the technological barriers to commercialization can be overcome so that the product can be taken to the market. Technology risk can be assessed as described in Table 6.2.

6.4.4.3 Invention Type

Companies typically consider certain invention types to be more valuable than others. For example, compositions of matter are traditionally understood to be the most valuable form of patent protection in the pharmaceutical industry because the protection afforded them is not generally limited by indication, route of administration, method of administration, or the like. In Table 6.3, platform technologies and compositions of matter both score 2, since both are extremely valuable.

Table 6.3 Evaluating invention type.

Rank	Score	Explanation
Platform	2	The invention is a platform technology
Composition of matter	2	The invention is a composition of matter
Method of treatment	1	The invention is a method of treating one or more specific disease conditions
Method of making	0	The invention is a method of making a composition of matter

6.4.5
Mapping the Competitive Patent Landscape

A critical component of any IP program is competitive IP intelligence. In the US alone, over 400 000 patent applications are filed each year.[7] Detailed patent studies must take into account not only patents of direct competitors but also patents of suppliers, customers, competitors of suppliers, and customers and companies in adjacent technology areas that may be relevant. The size and complexity of the task renders competitive IP intelligence most effective when it is accompanied by the development of a database for cataloguing and characterizing the large numbers of competitor patents.

Ongoing study of the patent literature in light of business strategy can help a company avoid the mistake of forging into a heavily patented area or at least permit the company to develop a rational strategy for working in heavily patented fields. In a recent example, a federal jury ordered Microsoft to pay $ 1.52 billion for infringing two Alcatel-Lucent patents for MP3 audio technology.[8] This amount is likely to be reduced on appeal, but it is easy to see that potential patent infringement damages are a significant risk for any company. Moreover, the patent literature of key competitors can provide insight into the competitors' priorities and strategies and can help to identify their strengths and weaknesses. Other uses of competitive IP intelligence include studies of patent portfolios of potential acquiring companies, technology licensors, and potential merger or acquisition targets. Figure 6.2 shows a simplified freedom-to-operate (FTO) process.

The phases of the process are typically increasingly costly and should be timed in a manner that is appropriate to a particular business. For example, searching and initial screening might be undertaken during initial evaluation of a product opportunity; mapping may be undertaken prior to initiation of development efforts; and FTO clearance might precede a first substantial investment in time and resources, such as scaling up for clinical studies.

---

7) http://www.uspto.gov/go/taf/us_stat.htm (last accessed 24 September).

8) Guth, R., Wingfield, N. "Microsoft Hit With $ 1.52 Billion Verdict in MP3 Suit." *The Wall Street Journal* 23 February 2007; p. A3.

FTO Process

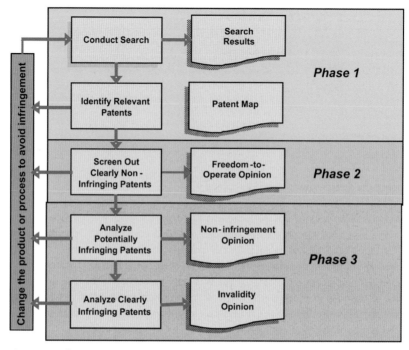

Fig. 6.2 Freedom-to-operate (FTO) process. The FTO process begins with the initial search and continues along increasingly detailed and complex levels of analysis. As the process progresses, risks are systematically identified and reduced using appropriate legal documentation and/or modifications to the company's product strategy.

The phased approach is also tailored to the fact that early in the product development process, the number of hypothetical products and infringement scenarios can be so great that serious infringement analyses becomes far too time-consuming and expensive. It is only when the number of potential product configurations is narrowed to a manageable number of alternatives that formal FTO analysis becomes practical. The level of detail of the FTO analysis must therefore be appropriately in sync with the development of clarity around the desired product design.

6.4.5.1 Conducting a Search

The FTO process begins with a patent search. It goes without saying that the quality of the rest of the process depends on the quality of the patent search. The purpose of the search is to identify all potentially relevant patents, i.e., every patent that could be a source of risk to the product development path. An effective search is both specific and broad and generally captures irrelevant patents that will be ruled out later in the process. The use of multiple search engines and multiple strategies (e.g., keyword searches, company searches, tracking down patents cited in key patents) can improve the comprehensiveness of the search.

There are many useful resources for patent searching, such as http://www.nerac.com, http://www.micropatent.com, http://www.derwent.com, http://www.delphion.com, http://www.uspto.gov and others. For example, to search for patents on somatostatin, the search term "somatostatin" should be entered. To narrow the search, the same search term should be used but only the patent's title, abstract, or claims searched. Students and others can often obtain free access to these services on a trial basis. Professional searchers are also available, and for some kinds of inventions, such as complex organic molecules, a manual search by a professional searcher is indispensable.

6.4.5.2 Identification of Relevant Patent Documents

Once the searching strategy has yielded a body of potentially relevant patents, the next step is to manually review them to identify relevant patents. This is typically a quick, high-throughput process in which the claims of each patent are reviewed to identify claims that could possibly encompass aspects of the product or methods of making or using the product. Patents with potential relevance should be retained for further analysis, and only those that are clearly irrelevant should be excluded.

6.4.5.3 Mapping of Relevant Patent Documents

Once the body of relevant patent documents has been identified, the documents can be visually mapped to reveal strategic information. Mapping involves the preparation of a variety of charts that permit visualization of patent data. The data being visualized are of two types: electronic data available from patent databases, such as the US Patent and Trademark Office's database, and data added by human analysis of the documents. An example of electronically available data is information such as the document's filing date, grant date, title, inventor, and the presence of various search terms. Human-added data include information such as technology categories, product categories, applications of the invention, and the like. An example of a patent map for patents related to somatostatin is shown in Figure 6.3. Figure 6.4 shows several charts made using similar data. Figure 6.5 shows a portion of the spreadsheet used to create the map and charts.

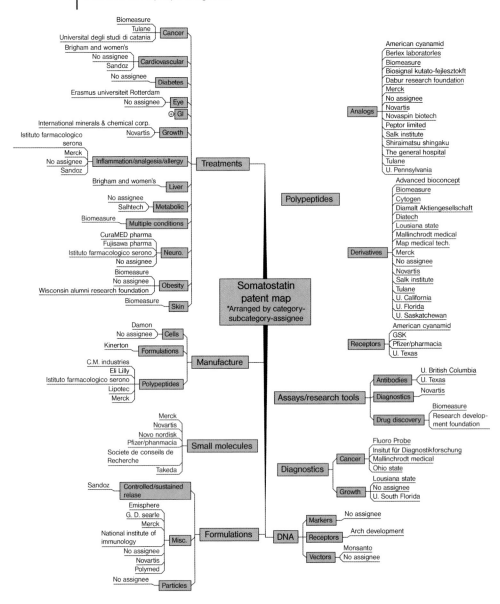

Fig. 6.3 Sample patent map for somatostatin. Somatostatin patents are "mapped" according to assignee, product and technology category, and indication.

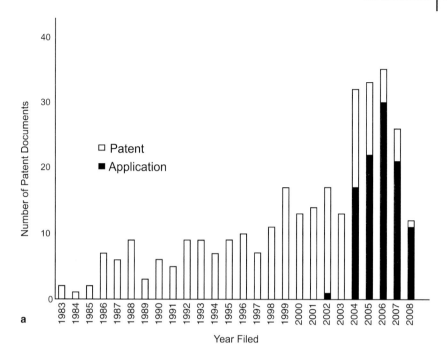

a

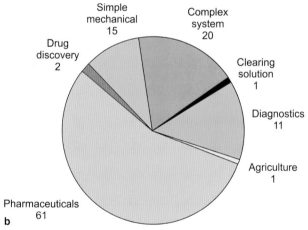

b

Fig. 6.4a–c Sample charts from somatostatin patent study. (a) Time analysis by year filed. This type of diagram visualizes the number of patents in the study increasing or decreasing. In this case the number of patents appears to be increasing or leveling off. Keep in mind that the 2005 data are only for a partial year. (b) Technologies of patents in the study. The primary technologies represented by each invention can be identified. Important strategic information such as the technology approaches used by competitors to solve specific problems can be gained from a chart like this. (c) Drug type by filing year. This chart shows the drug covered by the patent arranged by the year the application was filed. The strategic information gained from a chart like this could relate to industry trends or competitor interest in investing in particular technologies.

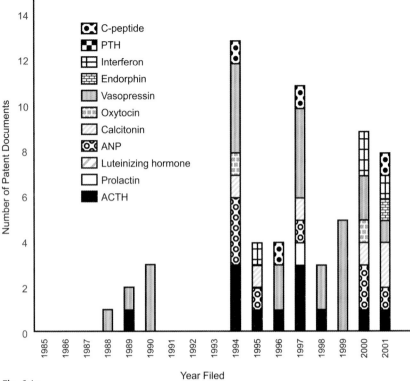

Fig. 6.4c

6.4.5.4 Screening Out Clearly Noninfringing Patents

In this step, the body of potentially relevant patents is reviewed in more detail. The company's patent counsel should review the claims and read the descriptive portions of the application to determine whether the claims would appear to be infringed by the product. This analysis will produce two sets of patents: one set of patents that will clearly not be infringed, and a second set of patents that are in a gray zone (i.e., it is still difficult to be certain whether or not they would infringe). For the first set of patents, the patent counsel can prepare an FTO opinion, which lists claims from the application, compares them to the proposed product, and explains why they will not be infringed. Patents in the gray zone should move to the next step for a more in-depth analysis.

6.4.5.5 Analyzing Potentially Infringing Patents

In this step, the level of detail in the analytical process becomes minute, as the company's patent counsel conducts a formal infringement analysis. This step requires the meaning of the patent claims to be carefully analyzed in light of the description of the invention in the patent specification and the file history of

Patent/Publication No.	Assignee	Title	Abstract	Category	Category 2	Brief Description
US20030031627A1	Mallinckrodt Medical	Internal image antibodies for optical methods and uses	Compositions and methods using	Antibodies		Internal image antibodies for
US5958790A	Nycomed Imaging	Solid phase transverse diffusion	The present invention relates to a	Assays/research tools	Analytical	Determination of an analyte in a test
US5460785A	RhoMed	Direct labeling of antibodies and other	Proteins containing one or more	Assays/research tools	Analytical	Labeling a protein containing
US5572025A	Johns Hopkins	Method and apparatus for	A method of operation of an ion	Assays/research tools	Analytical	Operation of an ion trap mass
US4855242A	Joslin Diabetes Center	Method of detecting antibodies	A method for determining the	Assays/research tools	Analytical	Determining the quantity of an
US20020132992A1	None	Reagent system and method for	Disclosed are a spectrofluorimetricall	Assays/research tools	Analytical	Analysis of a sample for a biologically
US4892813A	Centre National de la Recherche	Method for the determination of the	The invention relates to a method	Assays/research tools	Analytical	Measuring enzymatic activity of
US4764475A	U. British Columbia	Pancreas dependant	A correlation has been demonstrated	Assays/research tools	Antibodies	Method to identify a monoclonal antibody
US5998154A	U. Texas	Somatostatin receptor peptide	Peptide antigens derived from the	Assays/research tools	Antibodies	Antibodies from antigens derived
US20030224345A1	Advanced Cell Technology	Screening assays for identifying	The invention relates to assays for	Assays/research tools	Cells	Screening for stem cell research
US6610535B1	ES Cell International	Progenitor cells and methods and uses	The present invention relates to a	Assays/research tools	Cells	Pancreatic progenitor cells
US5861313A	Ontogeny	Method of isolating bile duct progenitor	The present invention relates to a	Assays/research tools	Cells	Isolating progenitor cells from a bile duct
US6271436B1	Texas A&M	Cells and methods for the generation of	Disclosed are methods for the	Assays/research tools	Cells	Growing porcine primordial germ
US5746996A	Immunomedics	Thiolation of peptides for	A method of radiolabeling a	Assays/research tools	Diagnostics	Method of labelling a binding agent for
US5753627A	Novartis	Use of certain complexed	Somatostatin peptides bearing at	Assays/research tools	Diagnostics	Somatostatin peptides with

Fig. 6.5 Sample spreadsheet of somatostatin data.

the patent. The file history is a record of the back-and-forth debate between the patent office and the applicant over whether the invention claimed in the patent application is patentable. File histories are usually hundreds and often thousands of pages in length.[9] If the conclusion of this infringement analysis is that the product does not infringe, the appropriate output is a noninfringement opinion clearly explaining the legal and technical rationale for this conclusion. A well-reasoned noninfringement opinion is required under US patent law to prevent increased damages, which may be awarded if the company is later sued for infringement of the patent and finds that a court disagrees with the conclusion.

6.4.5.6 Analyzing Clearly Infringing Patents

We can hope at this step in the process that there are no patents left; the analyses have established that the company's product strategy will not result in infringement of any patents, and, if necessary, the company's strategy has been modified to avoid infringing patents and/or the company has licensed or purchased necessary patent rights. However, if a patent remains that will clearly be infringed by the product and the company is unable to obtain a license, then the only remaining option is to show that the patent is invalid. This option is disfavored, in part because it is often quite difficult to demonstrate that a patent claim is invalid. Courts operate under the presumption that the patent office has done its job properly in issuing the patent, and this presumption can only be overcome by clear and convincing evidence of facts to the contrary. This legal standard places the responsibility of proof on one who would challenge the validity of a patent. Nevertheless, making a case for invalidity is often appropriate, especially where issued claims are unduly broad and prior publications can be found describing products or processes that fit within the broad claims.

Proper FTO management means doing the analyses necessary to safely embark on a product development path with minimal risk of infringing another company's patent(s). Many different FTO processes are possible; however, any process should include a search strategy for identifying all potentially relevant patents and progressing them through increasingly sophisticated analyses, appropriately timed to the product development process. Every patent entering the process should be analyzed and dealt with by making appropriate changes to the product strategy, acquiring ownership or license rights as needed and where appropriate, or by obtaining formal legal opinions explaining the basis of noninfringement or invalidity. It may be impossible to completely eliminate the risk of a patent infringement suit; however, with a working patent mapping and FTO process in place, the risk can be reduced and managed.

---

9) Many file histories for newer patents can be obtained from the US Patent and Trademark Office at http://portal.uspto.gov/ external/portal/pair (last accessed 16 September 2008).

6.4.6
Conclusion

Managing IP requires much more than just handing over invention disclosures to a patent attorney for eventual patenting. The process starts with a business strategy that employs IP as a key strategic tool. An IP vision will set the tone and focus IP decisions on the strategic target. Documenting a strategic IP plan will help to ensure alignment between the company's IP activities and its business objectives. A multidisciplinary IP team will help to ensure the development and execution of the IP plan. A disciplined invention assessment scheme will focus resources on the most valuable innovations. And a detailed understanding of the competitive patent landscape will inform the IP strategy and help prevent costly infringement problems.

6.5
Timeline

Session 1: Intellectual Property Law
Session 2: Patent Claims
Session 3: Intellectual Property Strategy
Session 4: Claims Mapping
Session 5: Patent Search
Session 6: Patent Databases and Patent Mapping
Session 7: Student Work
Session 8: Project Presentations

6.6
Study Plan and Assignments

6.6.1
Session 1

6.6.1.1 **Presentations**
Intellectual Property (Presentation #18)

6.6.1.2 **Assignment #1**
Read: Introductions to Chapter 5 and to this Chapter

Evaluate a Product or Service for IP Protection
Assemble an IP team. Assign each person a role: business, legal, and/or science. The business person will be primarily responsible for understanding business issues relevant to the company's IP. The legal person will be primarily

concerned with understanding legal issues relevant to the company's IP. The technology person will be primarily responsible for understanding the technological aspects of the company's IP.

Review the following article: Serguei P. Golovan, "Pigs expressing salivary phytase produce low-phosphorus manure," *Nature Biotechnology* 2001; 19, 741–45. Come to class ready to discuss the following questions about the article:

1. Technology
   (a) How does it work?
   (b) What problems does it solve?
   (c) What advantages does it have over existing technologies?
   (d) What is the existing state of the art and how is the invention novel relative to the state of the art?
2. Business
   (a) What would the product or service look like?
   (b) Is it a service or a product or both?
   (c) How would it be packaged, sold, used?
   (d) Who are the target customers?
   (e) Give the company's product or service a name that you will protect as a trademark.
3. Legal
   (a) What forms of IP protection will you use to protect the company's product or service?
   (b) What are the advantages of the forms of IP protection you have selected?
   (c) What are the legal risks associated with the forms of IP protection you have selected?

6.6.2
Session 2

6.6.2.1 **Presentations**
Discuss answers to the Session 1 questions and review the patent claims in US Patent 7115795 (which relates to the Golovan article reviewed for the assignment) to illustrate how the invention was protected. Map out the claims on the board to show how they are interrelated.

Claims and Infringement (Presentation #19)

6.6.2.2 **Assignment #2**
Review the file history of US Patent 7115795 – in particular, review the following documents:

10-23-2001   Claims (these are the claims originally filed in the application)
10-24-2001   Preliminary Amendment
10-24-2001   Applicant Arguments/Remarks Made in an Amendment
11-29-2004   Requirement for Restriction/Election

12-23-2004	Response to Election/Restriction Filed
12-23-2004	Applicant Arguments/Remarks Made in an Amendment
02-28-2005	Nonfinal Rejection
05-27-2005	Amendment – After Nonfinal Rejection
05-27-2005	Specification
05-27-2005	Claims
05-27-2005	Applicant Arguments/Remarks Made in an Amendment
05-27-2005	Authorization for Extension of Time All Replies
05-27-2005	Applicant Arguments/Remarks Made in an Amendment
08-12-2005	Nonfinal Rejection
10-11-2005	Amendment – After Nonfinal Rejection
10-11-2005	Claims
10-11-2005	Applicant Arguments/Remarks Made in an Amendment
01-12-2006	Final Rejection
03-22-2006	Amendment After Final Rejection
03-22-2006	Claims
03-22-2006	Applicant Arguments/Remarks Made in an Amendment
07-14-2006	Amendment after Notice of Allowance (Rule 312)
07-14-2006	Claims
07-14-2006	Applicant Arguments/Remarks Made in an Amendment

This file history records the debate between the patent examiner and the applicant seeking to patent the Golovan invention. Summarize the basic points made by the examiner in each of the rejections and the applicant's response to each of these points. Discuss how the claims of 23 October 2001 differ from the claims that were finally granted in the patent. What problems did the Golovan applicants run into that you would want to avoid in your patent application? What tactics did the Golovan applicants use to overcome these challenges?

6.6.3
Session 3

6.6.3.1 **Presentations**
Discuss the results of Assignment 2.
Managing Intellectual Property (Presentation #20)

6.6.3.2 **Assignment #3**

Evaluate a Product or Service for IP Protection
Identify another article from *Nature Biotechnology* with a technology for protection. The article must be less than one year old. Assemble your IP team. Pretend you are in a company, and the article is a brief describing new technology created by the company. Read the article and prepare a presentation answering the following questions:

1. Technology
   (a) How does it work?
   (b) What problems does it solve?
   (c) What advantages does it have over existing technologies?
   (d) What is the existing state of the art and how is the invention novel relative to the state of the art?
2. Business
   (a) What would the product or service look like?
   (b) Is it a service or a product or both?
   (c) How would it be packaged, sold, used?
   (d) Who are the target customers?
   (e) Give the company's product or service a name that you will protect as a trademark.
3. Legal
   (a) What forms of IP protection will you use to protect the company's product or service?
   (b) What are the advantages of the forms of IP protection you have selected?
   (c) What are the legal risks associated with the forms of IP protection you have selected?

Prepare a presentation introducing the technology, business, and legal aspects of the technology to the class.

6.6.4
Session 4

6.6.4.1 **Presentations**
Student presentations from previous assignment.

6.6.4.2 **Assignment #4**

Create a Claims Map
Conduct an IP team meeting in which you brainstorm to create a patent claims map for the company's technology.

Download the trial version of MindManager (http://www.mindjet.com/us/) or a similar mind mapping software. Other mind mapping software can be found at http://www.inspiration.com, http://www.visual-mind.com, and http://www.mindapp.com.

Create a claims map. Put the name of the product in the middle. Use a branching structure to show the area you will cover. If you are not sure about a specific topic, what question would you ask the company's scientist? Be creative. Think through the issues discussed in Assignment 1, such as protecting the invention in the form in which it will be sold. For example, if it will be sold as a kit, include claims to a kit. If it will be a bulk product, include claims covering the bulk product. Make sure to cover the following questions:

1. Who is the target market?
2. What problems does the invention solve?
3. How might someone engineer around the company's claims?
4. What additional research might help you to strengthen the company's patent protection?
5. Would the invention make a good trade secret?
6. What kinds of compositions of matter will you have?
7. Are there methods of using?
8. Are there methods of making?
9. Are there business processes?

Come to class with enough claims maps for everyone, and be ready to give a 15-minute presentation on your claims map.

Also, do a Google search on "Boolean searching." Review several Internet sites discussing this topic and come to class ready to work on some Boolean searches for your technology.

6.6.5
Session 5

6.6.5.1 Presentations
Presentations of claim maps.

6.6.5.2 Assignment #5

Patent Searching
Conduct a search of a patent database to identify patents of relevance to your technology.

Identify approximately 50 US patent documents (patents and applications). Use the US Patent Office Assignee database to identify current assignees. Create a spreadsheet with key information about the patents, including assignees, year filed, title, abstract, and patent number. Survey the patent documents and create at least five category columns to categorize the patents based on a scheme that is suitable to your technology area. For example, for a drug technology, columns might include: invention type (e.g., formulation, method of treatment, composition of matter, method of manufacture); disease condition (e.g., diabetes, inflammation, bacterial infection); route of administration (e.g., topical, oral, intravenous); formulation type (e.g., liquid, solid, cream). Use pivot tables to make a series of charts and maps visualizing the results of your analysis. Analyze the charts to identify strategic insights. Be ready to present the results of the study and, most importantly, be ready to describe the strategic insights gained from the study. Identify any problem patents and describe next steps in dealing with those patents. Identify any potential licensing opportunities.

6.6.6
Session 6

6.6.6.1 **Presentations**
Present work on Assignment #5

6.6.6.2 **Assignment #6**
Prepare a presentation describing your IP management plan. Describe the team and the roles of the team members. Include the technology, the market, the IP vision, the tactics you will use to protect the technology, your analysis of the competitive patent landscape, and how that analysis has impacted your strategy.

6.6.7
Session 7

Continue work on Assignment #6

6.6.8
Session 8

6.6.8.1 **Presentations**
Project presentations

7
Operational Excellence in Pharmaceutical Manufacturing

Lucia Clontz

Contents

Industry Immersion Learning. Real-Life Industry Case-Studies in Biotechnology and Business
L. Borbye, M. Stocum, A. Woodall, C. Pearce, E. Sale, W. Barrett, L. Clontz, A. Peterson, J. Shaeffer
Copyright © 2009 WILEY-VCH Verlag GmbH & Co. KGaA, Weinheim
ISBN: 978-3-527-32408-8

7.1
Mission

The mission is to introduce students to operational excellence using both concept- and laboratory-based applications in two case scenarios.

7.2
Goals

The goal is to provide students with knowledge of the regulations, demands, and technical challenges faced by management in the pharmaceutical and biopharmaceutical industries as well as an appreciation of the concept of operational excellence.

The team-based approach and real-world nature of the case scenarios will

- foster good communication skills;
- encourage innovative thinking;
- encourage the use of critical analysis to solve real-life challenges.

In addition, if the cases are conducted as a supervised practicum at a sponsor company, they will

- provide students with the opportunity to work in a "current good manufacturing practices" environment and gain new skills and knowledge;
- allow them to interact with and deliver to the sponsor company the product(s) of their work produced with quality and in a timely manner.

7.3
Introduction

7.3.1
Overview of the Drug Manufacturing Process

Pharmaceutical/biopharmaceutical manufacturing includes all operations involved in the process of making a drug product – from receipt and sampling of raw materials to the various unit operations in production, quality control, and, finally, packaging, labeling, storage, and distribution of the final product. All drug manufacturing companies must follow current good manufacturing practices (cGMPs), a set of regulations that are listed in the code of federal regulation (CFR) Parts 210 and 211 and in several Food and Drug Administration (FDA) guidance documents. cGMPs are defined by the FDA as "the minimum current good manufacturing practice for methods to be used in, and the facilities or controls to be used for, the manufacture, processing, packing, or holding of a drug to assure that such drug meets the requirements of the act as to safety, and has the identity and strength and meets the quality and purity characteristics that it purports or is represented to possess." Any operation that is part of a drug manufacturing process must be performed using approved standard operating procedures (SOPs) that detail the steps to be followed to ensure the quality and safety of the final product. Any deviation from established procedures must be investigated and explained. Changes to an approved drug production process must be thoroughly tested, explained, and justified to the regulatory agencies. In this climate of great scrutiny and adherence to standards, a company must be profitable and efficient while delivering the quality and safe products customers expect. Therefore, companies have established management systems to monitor the various operational steps in an effort to identify opportunities for system optimization and improvement.

7.3.2
A Change in Paradigm for the Pharmaceutical Industry

In the pharmaceutical/biopharmaceutical industry, meeting standards of safety and quality in drug manufacturing is an important focus of companies. Over the years, drug manufacturers have encountered internal pressures to improve operational systems and increase profits and also regulatory pressures to design quality into processes. Companies are embracing a concept known as "operational excellence" in an attempt to increase profit margins by gaining efficiencies in operations and reducing waste. Operational excellence can be defined as a business process or program that identifies opportunities for improvement. Examples of such opportunities include improvements in documentation systems, meeting effectiveness, better planning, communication and coordination goals between departments, as well as changes in how production is managed, such as optimization of operation systems in manufacturing and reductions in cost and production cycle times. In summary, operational excellence is about a company truly understanding its internal operating standards, processes, and capabilities so that continuous improvement initiatives can be implemented and monitored, resulting in an efficient and profitable organization that ensures the quality of goods and services.

In the regulatory arena, the FDA has been the main driving force for improvement in operations. In August 2002, the FDA launched an initiative called *Pharmaceutical Current Good Manufacturing Practices (cGMPs) for the 21st Century.*[1] This initiative was designed to enhance and modernize the regulation of pharmaceutical/biopharmaceutical manufacturing and product quality. Among the focal points of this FDA initiative are the following:

- Implementation of process-understanding principles – designed to ensure that a company has the technical understanding of its manufacturing pro-

1) *Pharmaceutical cGMPs for the 21st Century (2004) – A Risk-Based Approach, Final Report.* Department of Health and Human Services, US Food and Drug Administration, Rockville, MD, 2004.

cesses to be able to adequately and proactively implement systems to measure (metrics) and control the processes, with the ultimate goal of reducing variability and ensuring product quality and safety.

- Risk-based approaches in the manufacturing industry – designed to ensure that a company understands the risks associated with process changes so that risk to product quality and safety can be mitigated.
- Risk-based approaches during regulatory inspections – to be implemented due to limited resources within the FDA. The new regulatory inspection approach adopted by the FDA utilizes more effectively its available resources by prioritizing sites for inspections based on low- vs. high-risk operations. Operation designation could be based on company history with the FDA as well as the type of product manufactured (e.g., sterile vs. nonsterile products).
- Decision making based on state-of-the-art pharmaceutical science – designed to ensure that a company has the scientific understanding needed to make process changes and to implement process controls that will result in satisfactory product performance. The FDA's new inspection approach takes advantage of emerging science and data analysis to ensure efficiency and effectiveness of audits.
- Early adoption of new technological advances – companies are encouraged to implement alternative and rapid methods as part of the process analytical technology (PAT) initiative, which can provide benefits such as real-time results and timely assessments of potential process deviations.

Indeed, the FDA has provided for change in the pharmaceutical industry, a complete shift to a focus on process understanding and implementation of new technologies.

The following two case scenarios were designed to illustrate how operational excellence initiatives can be applied as a business management tool to address efficiency gains and as a means to introduce new and improved scientific methodologies in operations.

7.4
Part I – Operational Excellence: Implementing Process Improvements

7.4.1
Introduction to Lean Manufacturing

"Lean manufacturing" is a comprehensive term referring to manufacturing methodologies based on maximizing *value* and minimizing *waste* in a manufacturing process. Value is defined by the customer – either internal to the company (e.g., manufacturing can be viewed as a customer of quality control laboratories and, therefore, to manufacturing, value is often a fast turnaround and reliability in test results) or external (e.g., a client to whom value is a quality product delivered on time). Waste can be defined as steps in a process that are not needed or are redundant and as such do not add value.

The evolution of lean manufacturing began in the automotive industry with the Toyota Production System (TPS) in Japan. [2] Many of the most recognizable terms used in lean manufacturing, including *kaizen* (continuous improvement; persistence to eliminate unprofitable situations), *andon* (a type of alarm system to show different status conditions), and *kanban* (a system, often with cards, that facilitates just-in-time delivery), are Japanese terms.

Lean manufacturing, along with "six sigma concepts," which are discussed later in this chapter, has become the leading business process tool in today's successful manufacturing organizations. Lean manufacturing principles provide a way for companies to produce more with less human effort, less equipment, less time, and less space while providing customers with exactly what they want. Thus, a lean manufacturing company strives for an ideal operational environment that makes products and provides services of the highest quality and fastest delivery and with the lowest cost to the company. The following are examples of situations in a company that is not considered lean:

- Fire fighting
- Poor financial performance
- Underutilization of employees
- Inconsistent policies
- Too much paperwork
- Too many handoffs (too many people involved)
- Poor training
- Lack of team work
- Frequent or excessive downtime (time when production is at a halt)
- Frequent and excessive re-work or rejects/recalls (unacceptable products)

The key lean manufacturing principles are:

- Delivery of value ensuring that the company meets customers' (internal and external) expectations
- Improved flow of processes in an attempt to streamline and standardize work flows to achieve reduced cycle times and eliminate potential situations that can negatively impact operations
- Reduction of waste (the inverse of value creation) that can negatively impact the work flow or incur unnecessary costs to an operation

Two ways to evaluate opportunities for improvement and waste reduction are:

- Process mapping, often referred to as "brown paper sessions" since many companies draw a flow diagram of the process on a large brown paper pasted to a wall in a conference room and hold meetings with key individuals to get feedback (positive and negative) so they can identify gaps (missing essential components, tasks, etc.), critical steps, and improvement opportunities.

2) *A Brief History of Lean,* Lean Enterprise Institute. http://www.lean.org/WhatsLean/ History.cfm (last accessed 24 September 2008).

- Time and motion studies, which can be defined as a systematic way to evaluate the time and human motions during the performance of an operation so that waste steps can be identified. This type of study can be a key source of productivity improvements. For example, a time and motion study analyzing the time it takes and the motions executed to process a sample in a quality control testing laboratory may be able to identify savings in material costs and processing time.

Other key principles in lean manufacturing include roles, responsibilities, and culture, and alignment of departmental goals. Professional development of employees and building of work relationships go hand-in-hand with the success of operations. Therefore, most operational excellence initiatives include programs in people development and team building. This includes initiatives to improve the company training program in areas such as delivery of training, training effectiveness, leadership development, team work, human error reduction, and communication tools.

In the 1980s, Motorola worked at implementing systems that would reduce the number of defects in their products and called the program Six Sigma. [3] Since sigma is a statistical measure of variability, Six Sigma initiatives strive to reduce variability in manufacturing to a level of almost perfection/freedom from defect (e.g., 3.4 defects per million opportunities or DPMO). The standard for most companies is to operate at a Four Sigma level (e.g., 6000 DPMO).

Key Six Sigma principles are:

- Reduction of total defects
- Increased capacity and output
- Products/services of better quality and reliability
- Better financial results
- Increased customer satisfaction

A typical process followed when using Six Sigma principles is called DMAIC – *define, measure, analyze, improve,* and *control*.

- *Define* – ensures the company understands what is critical to quality so that the problem and action plan can be defined.
- *Measure* – ensures a company focuses on measuring the critical and relevant aspects of the process, also known as key performance indicators (KPIs).
- *Analyze* – ensures a company identifies the factors that are key to driving performance.
- *Improve* – ensures a company establishes improvement goals based on KPIs.
- *Control* – ensures a company defines a plan to control process improvements and assures sustainability of those improvements.

A typical tool used in Six Sigma is the Pareto chart. This tool, developed by Wilfredo Pareto, shows that in any business the greatest improvement is achieved

3) Six Sigma Executive Overview, North Carolina State University, College of Textiles, Industrial Extension Service, Raleigh, NC, 2004.

when people focus on 20% of the issues that cause 80% of the problems. Other tools include process flow diagrams/process mapping, statistical process control charts, brainstorming sessions, and cause-and-effect/fishbone diagrams.

Traditionally, the pharmaceutical industry has not incorporated lean manufacturing or Six Sigma principles into its infrastructures since such concepts may seem difficult to apply in a highly regulated environment. In addition, given the fact that operational excellence initiatives often involve changes and most people are skeptical or do not like changes to their routine, management is left with the task to implement not only process and system changes but also a culture change in an attempt to make people become more receptive to new and improved ways of doing work. The best way to achieve such culture change is to show results that will benefit the people in one way or another. Companies can start continuous improvement initiatives by looking for the "low-hanging fruit" and "quick wins" to help shift the paradigm in company culture. Indeed, *successful change programs focus on results*. Drug manufacturers who have decided to adopt the lean manufacturing and Six Sigma models and have focused on a results-driven transformation have demonstrated that this concept works well in the pharmaceutical manufacturing environment.

7.5
Predicted Learning Outcomes

Gaining knowledge in process improvement techniques and process understanding is an invaluable benefit to students who are preparing themselves to enter the work force. This case was designed to demonstrate how lean manufacturing tools can be applied in a pharmaceutical company in order to achieve process improvement. Specifically, students will learn

- how to differentiate value-added from non-value-added activities;
- how to apply process mapping techniques;
- how to perform time and motion (T&M) studies used to analyze the efficiency with which an industrial operation is performed (applicable to a company-sponsored practicum);
- how to assess the level and effectiveness of planning and communication that exists among certain departments that could lead to identification of constraints and cause unnecessary waste;
- how to prepare service level agreements (SLAs) used to align goals and define roles and responsibilities between departments.

In summary, using lean-manufacturing tools, students will learn how to identify, recognize, and define problems that would prevent the company from achieving the desired value. Then, students will be able to characterize, measure, and analyze the problems and implement procedures to mitigate them, thus creating value.

7.6
Case Scenario

This case can be designed to be completed in an eight-week period. The instructor may follow the guidelines given in this chapter in order to organize the project. Students are expected to dedicate a minimum of 4–10 hours a week to the project. The instructor can create a fictitious company, describe its various business challenges and choose one or more areas that are in need of process improvement. The following case scenario illustrates a typical improvement opportunity at a pharmaceutical company:

> The manager of the Facilities and Engineering department (responsible for plant and equipment maintenance, repairs, and calibration) at a pharmaceutical manufacturing company needs help with improving the work order process (the procedure used by employees to request services from the department). His internal customers, mainly manufacturing and quality control personnel, have complained of lack of planning for plant and equipment maintenance activities as well as delays in completing scheduled work. The current work order process is not integrated with planning of activities in other departments, leading to unnecessary delays. In addition, although the company has a computerized system to manage and track maintenance, repair, and calibration work performed, employees still have to use a paper-based system and fill out a service request form (SRF) whenever they need service from Facilities and Engineering. Indeed, the company is in need of a management solution that will optimize the work order process to benefit various parts of the organization (internal customers).

7.7
Timeline

Session 1: Operational Excellence, Lean Manufacturing
Session 2: Drug Development, Regulations
Session 3: Areas for Improvement, Process Mapping
Session 4: Student Work
Session 5: Service Level Agreements
Session 6: Student Work
Session 7: Student Work
Session 8: Student Presentations

7.8
Study Plan and Assignments

7.8.1
Session 1

7.8.1.1 Presentations

Review the basics about the principles and tools of operational excellence and lean manufacturing.

Steps to Lean Manufacturing (Presentation #21):

- Overview of operational excellence
- Overview of lean manufacturing and Six Sigma
- Define value-added and non-value-added steps in a process
- Discuss process mapping principles
- Provide examples of process flow diagrams

7.8.1.2 Assignment #1

Read: Introduction to Part I

Create and Improve a Familiar Process

Before working on the proposed scenario, students must learn about lean manufacturing principles and practice creating process maps for identification of value-added and non-value-added steps in the process. The instructor should select a familiar process for the exercise to which students can relate. The following example can be used by the instructor to introduce the concept of lean manufacturing and lean principles.

Example Assignment: Process to make chocolate chip cookies

1. Divide students into three teams that will represent three departments in a company that makes chocolate chip cookies.
2. Team 1 will define the requirements for baking a batch of chocolate chip cookies in terms of processing and delivery times.
3. Team 2 will create a list of materials, number of personnel, and equipment needed to make the cookies.
4. Team 3 will write a detailed recipe to make the cookies, including equipment requirements.
5. At the end of this exercise, the three teams will meet to discuss the project and compare notes.
6. Team members will create a process flow diagram (process map) for the project and compare the timeline from the process map with the timeline provided by Team 1.
7. If the mapped process will take longer than the timeframe specified by Team 1, students must identify which steps in the process are critical, which ones add value to the overall process, and which ones can be eliminated or combined to reduce the process timeline.

8. If the mapped process meets the time requirements from Team 1, the instructor should intervene and play the role of plant manager and require that the process be improved, for example, either in terms of time, number of personnel, or materials. This will ensure that students will have a target to work from so they can start to understand the concept of process improvement.
9. After improvement opportunities are identified, students should revise the process flow diagram, which should be the deliverable for the exercise.

7.8.2
Session 2

7.8.2.1 Presentations
The instructor presents the scenario for work process improvement. Since this case illustrates a process in a pharmaceutical company, students must learn about the FDA and standard operating procedures.

Drug Development: Regulatory Bodies and Requirements (Presentation #22):

- Definition of cGMPs
- Overview of the FDA
- Overview of the FDA 21 Code of Federal Regulations (CFR), Parts 210 and 211
- Overview of FDA initiatives for the 21st century
- Discuss content and role of SOPs in a regulated industry

The instructor can elect to create study questions designed to encourage discussions on the topics being addressed in the case, to include theory presented in a prior lecture. These questions can range from technical to business- or personnel-related issues. Typically, these study questions provide supplemental material related to the topic of the case that perhaps will not necessarily be presented. If they are critical to understanding the challenges faced by pharmaceutical/biopharmaceutical companies they must be taken into account when proposing the improvements for the process.

7.8.2.2 Assignment #2

Improve the Work Order Process at a Pharmaceutical Company
After presenting the scenario, students should be divided into three teams that will represent key departments in the company that are involved with the work order process: Facilities and Engineering (the service provider), Manufacturing (the customer), and Planning (the group that plans and schedules the work).

The instructor will play the role of a plant manager and present the current process flow in a diagram as seen in Figure 7.1 and highlight some of his or her concerns. The flow diagram identifies the waste steps and bottlenecks in the process and includes recommendations for process improvement. These should be left out of the diagram that will be presented, since the assignment is for the students to identify improvement opportunities and make recommendations.

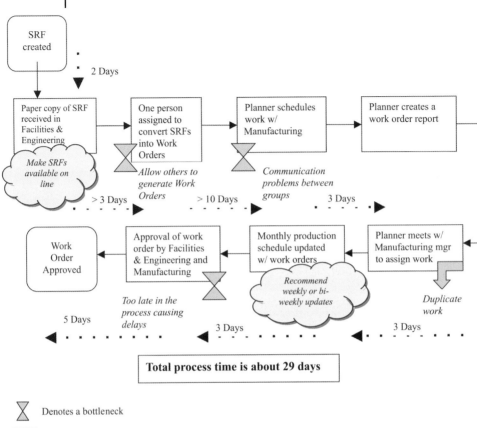

Total process time is about 29 days

 Denotes a bottleneck

Denotes a waste/non-value-added step

Denotes recommendations

Fig. 7.1 Summary of work order process.

The instructor can summarize the work order process as follows:

1. Service identified by customer (Manufacturing). Service could be calibration of a piece of equipment, for example.
2. Service request form (SRF) created in paper and mailed interoffice to Facilities and Engineering department.
3. Billy Jean is in charge of converting SRFs into work orders in paper. She sends the paperwork interoffice to Joe Doe, the company planner.
4. Joe Doe enters the information in a spreadsheet and contacts Manufacturing by e-mail to schedule the work.
5. Joe Doe plans a meeting with the manufacturing manager to schedule the work.

6. Joe Doe updates the monthly production schedule to include the needed work for the piece of equipment.
7. After the manufacturing manager agrees with the schedule for the work to be performed, Facilities and Engineering approves the work.
8. Total timeframe from identification to approval of work needed is 29 days.

7.8.2.3 Assignment #3

- Form the three teams and start gathering information on the process being evaluated.
- The team representing Facilities and Engineering will start writing an SOP to describe the work order process.
- The team representing Manufacturing will decide on the work that needs to be performed and specify, in writing, their needs in terms of service, including turnaround time.
- The team representing Planning will start evaluating improvement opportunities in their functional area.

7.8.3
Session 3

7.8.3.1 Presentations

The instructor reviews with students the progress made from Session 2 and presents additional information that will help them with their assignment.

Identification of Areas for Improvement (Presentation #23):

- Discuss the major challenges faced by an organization that is trying to improve a process that is cross-functional (e.g., involves/impacts several departments in the organization).
- Provide an example of time and motion study concepts and examples (applicable to a company sponsored practicum).

7.8.3.2 Assignment #4

Students will continue working on their assignments as follows:

- Team representing Facilities and Engineering will distribute their work order process SOP to the other groups for review.
- Team representing Manufacturing will present their expectations for the work order process to the other groups.
- All three groups will start a brown paper session (process mapping) to evaluate improvement opportunities. It is expected that once the flow diagram is created on the brown paper, students representing the various company functional departments will give their input for improvements, which can be added to the process map using sticky notes that can be easily moved around.

7.8.4
Session 4

7.8.4.1 Presentations
The instructor will discuss the progress of the assignment and present his or her expectation for an interim team report. This document must include preliminary identified improvement opportunities and should include approximate cost for any equipment or software that needs to be purchased and any additional head count (employees).

7.8.4.2 Assignment #5
Students are expected to continue working on the process mapping exercise and start making decisions for the final recommendations for improvement. The deliverable at this stage of the project is an interim report that should be due the following class. This report is a critical deliverable in case of unexpected situations that will prevent the project from being completed in the agreed timeframe or in case an alternate path must be taken. The interim report must include the expected cost that the company will have to incur to meet the target timeline for the work order process in the form of a cost vs. benefit analysis.

7.8.5
Session 5

7.8.5.1 Presentations
The instructor will discuss the concept of service level agreements (SLAs) and present an example as seen in Figure 7.2. The instructor will also review with students the interim reports and as plant manager, will decide whether the proposals for improvement are a "Go" or "No Go."

Service Level Agreements (Presentation #24)

7.8.5.2 Assignment #6
The students will evaluate the feedback received from the instructor and start preparing for the revision of their documents. The following are examples of deliverables at this stage of the project.

- The team representing Facilities and Engineering will revise their work order process SOP to reflect improvement changes.
- The team representing Planning will write a proposal to the plant manager for improvements needed in their functional area.
- The team representing Manufacturing (the customer) will start drafting a service level agreement between the Manufacturing department and the Facilities and Engineering department.

Service Level Agreement
Facilities and Engineering (F&E) has committed to timely evaluation and execution of service and maintenance work based on service requests forms (SRFs) received by Manufacturing. Manufacturing has committed to providing sufficient notification for work order requests to F&E whenever possible, with the exception of emergency work.

Facilities and Engineering	**Manufacturing**
☐ Manages and plans work order requests, so that production schedule is not negatively impacted. ☐ Evaluates assigned SRFs. ☐ Maintains an updated list of manufacturing equipment with calibration and preventative maintenance (PM) due dates. ☐ Performs commissioning and validation activities ☐ Responsible for creating equipment specific SOPs.	☐ Submits SRFs for new equipment and other service request needs for facility and production equipment. ☐ Follows equipment procurement SOP for new purchases. ☐ Provides list of equipment in manufacturing suites. ☐ Provides timeline for service in agreement with F&E ☐ Provides support and coordination for execution of equipment validation. ☐ Provides support for creation of equipment SOPs.
Turnaround Time ☐ Responds to service request forms within 3 working days. Approves service request forms within 30 working days of receipt. ☐ Completes equipment specifications within 30 working days from request. ☐ Ensures that electronic updates of equipment drawings will be made available for approval within 15 working days from request. ☐ Commits to execution of equipment qualification based on project schedule.	Turnaround Time ☐ Submits service request forms for all equipment services and project engineering requests. ☐ Provides feedback on engineering documentation within 7 working days from receipt of document. ☐ Provides review and approval of equipment drawings within 5 working days from receipt of electronic documents. ☐ Provides assistance to execution of equipment qualification based on project schedule.
Communication ☐ Notifies Manufacturing within 15 days of scheduled/planned PM service.	Communication ☐ Provides F&E notice of new equipment within one week after purchase approval.
Requirements/Constraints ☐ Agreement only applies to scheduled work orders. ☐ Agreement only applies to 1st and 2nd shifts ☐ No routine 3rd shift or weekend PM support. ☐ Emergency service and off-shift maintenance and repair work are addressed via the on-call service.	Requirements/Constraints ☐ Manufacturing must coordinate with F&E for work performed to prevent significant production downtime. ☐ Manufacturing must use the on-call service for off-shift maintenance and repair needs.
SLA Contacts Print Name/Signature/Date	SLA Contacts Print Name/Signature/Date
Metrics to be Used to Monitor Agreement Facilities and Engineering will implement a tracking system to measure the attainment of service target turnaround times. Manufacturing will implement a tracking system to measure the attainment of target timely service received by Facilities and Engineering. Metrics will be provided to the SLA contacts on a monthly basis.	

Fig. 7.2 Example of a service level agreement (SLA) between the Facilities and Engineering department and the Manufacturing department.

7.8.6
Session 6

7.8.6.1 Presentations
The instructor should plan the lecture for Session 6 in a company meeting style. In this meeting, the instructor, playing the role of plant manager, will review the proposed revised process map, the proposed revisions to the work order process SOP, and the proposed SLA. At the end of the meeting, the instructor must ensure that consensus is reached among the three teams.

7.8.6.2 Assignment #7
Students are expected to actively participate in the meeting as if they were representing their functional areas in a real company. After consensus has been reached, students should join their groups and start working on the final case deliverables as follows:

- The team representing Facilities and Engineering will finalize their SOP for the work order process.
- The team representing Manufacturing will finalize the SLA between the Manufacturing department and the Facilities and Engineering department.
- The team representing the Planning department will create a new process flow diagram that includes the improvements agreed upon. Refer to Figure 7.1 for an example of how the diagram should be designed.

7.8.7
Session 7

7.8.7.1 Presentations
The instructor provides any additional support needed to ensure completion of the project in the expected timeframe. The instructor should also provide clear expectations for the final team reports that will summarize the various assignments conducted by the three teams of students.

7.8.7.2 Assignment #8
Students complete their deliverables and start working on their final team report and a PowerPoint presentation for the last day of class. The team report must contain the old and the improved process as well as benefits/gains to the company.

7.8.8
Session 8

Instructor Tasks: This is the last day of class when the students present their work to the instructor and their classmates. The instructor must ensure provisions are made ahead of time for this event. Some of the recommended tasks for the instructor are:

- Coordination of logistics for the presentations
- Invitation of guests to the event
- Preparation of questions for the groups to test their knowledge of the subjects covered during the course of the semester
- Evaluation of the groups' presentations based on content, innovation, and skills
- Ensuring that questions are posed to the groups to test their ability to answer questions posed by an audience

7.8.8.1 Assignment #9
This is the day when the project ends and the students will be evaluated on their deliverables. Student assignments for the last day of class should include:

- Delivery of all the work collected and prepared, which must include a written final report and proposal for improvement
- Summary of work performed in the form of a PowerPoint (slide show) presentation
- Preparations to answer questions posed by audience and instructor

7.9
Company Supervised Practicum

This case can be conducted in a classroom setting as detailed in this chapter, or as a supervised practicum at a real company. Many business processes that exist in a pharmaceutical company can be adapted and developed into classroom team projects.

If the case is carried out as a supervised practicum at a pharmaceutical/biopharmaceutical company, several preparations must take place ahead of the study. In particular, when a case is carried out in a cGMP-regulated environment, provisions must be made for students to receive the training required (e.g., cGMP training, safety, documentation practices, etc.). Students must also sign a confidential disclosure agreement (CDA) before they can actually carry out any work at the company. A CDA requires the recipient of sensitive company information (e.g., protocols and process descriptions) to keep the information in confidence. Figure 7.3 contains a checklist that the instructor can use to help prepare for a company sponsored practicum.

☐ Confidentiality Disclosure Agreement (CDA) between students and company

☐ Background check performed for students

☐ Creation of a project plan with clear deliverables and timeline

☐ Badge access and photos

☐ Desk/work space for students

☐ Office supplies for team

☐ Phone access for team

☐ Computer access for students

☐ Company training scheduled (e.g., cGMP training, documentation practices, safety)

☐ Laboratory access, personal protective equipment (PPE), and safety training for laboratory-based case studies

☐ Materials and equipment available for laboratory-based case studies

☐ List of key company contacts (names and phone numbers)

☐ Weekly meeting with student team leader to monitor progress of project

☐ Key company employees invited to students' presentations at project completion

☐ Meeting room reserved for students' presentations

Fig. 7.3 Check list of key activities for a company supervised practicum.

The training needed in industrial settings may delay the actual project by two to three weeks. Therefore, it is imperative for the instructor to include the company-required training in the project timeline. Optimally, students could prepare for the project before undergoing company training with preproject reading assignments. These could include information on lean manufacturing, process mapping, time and motion studies, and related procedures for the chosen process. For example, if a work order process is involved, company SOPs and flow diagrams (if available) describing the process that needs to be evaluated could be included in preproject study material.

A case similar to the one described in this chapter was carried out as a supervised practicum in a biopharmaceutical company by a team of four students, who were supervised by a company employee, the manager of Facilities and Engineering. Figure 7.4 contains the abbreviated project plan that was created by the company manager. This document formed the basis for the work that was performed by the students.

The students who participated in this project were required to undergo a total of eight hours of training in cGMPs, company safety, and documentation practices. Students spent at least three hours a week at the sponsor site. However, they were expected to dedicate about 10 hours a week to the project. The time outside the company was used to review and create documents and process flow diagrams as well as for brainstorming sessions. On site, students spent their time in meetings with industry professionals, setting up brown paper sessions, performing time and motion studies, and interviewing employees.

Work Order Process Improvement Project Plan
Prepared by Bill Smith
Manager, Facilities and Engineering

Purpose: To document the existing process flow for maintenance work orders with activities for the various functions within the organization and to identify opportunities for improvement.

Goal: To identify the non-value-added activities related to the work order process and make recommendations for improvements so that a higher level of efficiency is achieved with the existing work force and the processing time is reduced to less than two weeks.

Scope: This project will involve the maintenance department functions that support the quality control laboratories and manufacturing areas.

Deliverables: The main deliverable is a ***Report*** that identifies the current work process flow (process map/process flow diagram), which includes the non-value-added activities. This report should also include recommendations for process improvement and an improved process flow diagram.

Outcomes: The project report will be used to revise existing standard operating procedures that define the work order process.

Milestones:
- Week One: Project begins; company access (badge, photos, etc.), company training and general introduction to project and key company contacts.
- Week Two: Company training and review of company documents
- Weeks Three–Five: Begin interviews of employees (key contacts and internal customers)
- Week Six: Review results prior to report
- Week Seven: Final input and revisions
- Week Eight: Presentation to company management and delivery of final product (deliverables)

Fig. 7.4 Example of an abbreviated project plan for a business process case.

The students adhered to the schedule provided to them by the company project leader and were willing to go to the company on days other than the scheduled class day in order to meet with their appointed company contacts. During the course of the semester, the students interviewed several company employees, including maintenance technicians and manufacturing managers. They also spent time shadowing maintenance personnel as part of the time and motion studies in order to collect suggestions, identify non-value-added steps, and detect problems faced by the employees during the course of their daily work. After each interview and after each time and motion study event, students updated the existing process flow map.

The project at the sponsor company was successfully completed within the eight-week period, and all deliverables were provided to the company project leader as planned. The students were able to decrease some of the "silo mentality"(lack of communication and common goals between departments) prevalent at the company and make practical recommendations for process improvements. Some of the suggestions made included on-line access to SRFs, job task sharing to avoid bottlenecks, removal of the waste steps in the process, and creation of SLAs between key functional departments. If all recommendations were implemented, the total process time for work orders could be reduced from an average of 30 days to 15 days without additional staff or automation.

According to the students, the main factor that contributed to the success of the project was having regular update meetings with the Facilities and Engineering manager, who served as the project leader. Indeed, company project leaders play a critical role in the success of case studies. These individuals must be available to help the team develop the right solutions so that students do not waste time making recommendations that go against company policies or that are economically prohibitive. In addition, having scheduled interviews with the various company employees helped maximize the students' time at the sponsor site; it is very important for students to realize that they should not waste valuable company resources. Therefore, time management is critical to ensure that issues that surface during the course of the project are addressed in a timely manner.

One student commented that, in the beginning, he was afraid that the team's recommendations would not be well received because they had experienced some resistance during the interviews. Besides, he knew that most people resist change and from the comments he received, discerned that company employees were no exception. However, by the completion of the project, the team realized that their suggestions were not only welcomed by management, but that management was actually seriously considering revising company procedures to incorporate the students' recommendations. The present author was able to observe a great sense of empowerment in the students who participated in this case; students realized that by using lean manufacturing tools, they were able to contribute to significant improvement changes in a real pharmaceutical company.

7.10
Part II – Optimizing Existing Technologies

7.10.1
Introduction to Quality Control

The Quality Control (QC) group plays a critical role in pharmaceutical and bio-pharmaceutical manufacturing. Test results generated confirm the purity, identity, potency, stability, and safety of drug products. In fact, a significant portion of the cGMP regulations deal specifically with QC laboratories and product testing. It is a regulatory requirement that all QC testing be performed in a manner to ensure the validity and accuracy of the test results generated. Therefore, method qualification and validation and the use of test controls are common practices in a pharmaceutical QC laboratory. The QC group also performs testing to ensure the suitability of equipment and facilities for use in product manufacturing (e.g., environmental monitoring, cleaning validation, disinfectant testing). These types of production support testing often require some type of method qualification.

While most operational excellence initiatives focus on business processes, process understanding and continuous improvement efforts can also be implemented for technology-based applications to reduce product variability, increase product yields, and improve test methodologies. Many of the drivers for testing method improvements come from risk assessments, defined as a process of systematic and objective evaluation of available information pertaining to the risk of failure (including contamination) of a product, system, or method. Based on the results obtained from the risk assessment, the company uses risk management to weigh the possible alternatives to select and implement appropriate controls to mitigate risk.

Using concepts of science-based risk management as defined by the FDA, a different approach to operational excellence is taken with this case; students are assigned a laboratory-based project to develop and implement a new methodology to reduce the risk of failure of a process or product due to equipment contamination with biofilms.

7.11
Predicted Learning Outcomes

This case was designed to illustrate the application of process improvement initiatives to QC testing in pharmaceutical manufacturing. Students will learn about

- the role of the FDA in the pharmaceutical/biopharmaceutical industry;
- the role of quality control, quality assurance, manufacturing sciences and cleaning validation, and regulation in pharmaceutical/biopharmaceutical manufacturing;

- the principles of method and equipment qualification and validation;
- the United States Pharmacopeia (USP);
- the FDA initiative on science-based risk management;
- documentation in a cGMP environment.

7.12
Case Scenario

This science-based case can be designed to be completed in an eight-week period. The instructor may follow the guidelines given in this chapter in order to organize the project. Students are expected to dedicate four to 10 hours per week to the project.

The example case given in this chapter addresses biofilms in a biopharmaceutical manufacturing environment and involves qualification of equipment sanitization procedures. Biofilms are complex and three-dimensional microbial sessile communities of cells embedded in a matrix of extra polymeric substances (EPS) and irreversibly attached to a substrate, interface, or each other. [4]

Biofilms are universal and found to be a preferred way of life for bacteria; biofilms represent a multilayer defense mechanism and a survival strategy against environmental stresses. Once microorganisms attach to a surface and form a biofilm, the organisms will undergo a complete change in behavior and phenotype, making them much more resistant to typical chemical sanitizers that kill their planktonic counterparts. This is a problem because methods for qualification of sanitizers and disinfectants usually are carried out using planktonic cells and not biofilm cells.

Microbial biofilms are of growing concern to pharmaceutical and biopharmaceutical manufacturers. Biofilms present in water systems and production equipment can adversely affect the quality and safety of products and cost companies thousands to millions of dollars annually in equipment damage, production downtime, lost batches, investigations, and energy losses.

The following case scenario illustrates a typical method improvement opportunity at a biopharmaceutical company.

> The plant manager at a biopharmaceutical company needs to develop alternate and improved methods for cleaning chromatography columns that have become contaminated with bacterial biofilms and for testing the antimicrobial properties of sanitizers and disinfectants used to clean biofilm cells from equipment. The manager understands the limitations of traditional test protocols for evaluating the antimicrobial efficacy of sanitizers, since these methods use planktonic cells of representative test organisms that might be present in a pharmaceutical manufacturing environment. Therefore, the manager wants the help of a team of scientists to rethink

4) Donlan, R. M., Costerton J. W. "Biofilms: survival mechanisms of clinically relevant microorganisms." *Clinical Microbiology Reviews* 2002; 15(2): 167–193.

the traditional protocols for equipment cleaning. In addition, the manager wants the scientists to develop a method that uses biofilms grown in a laboratory setting for use in sanitizer and disinfectant antimicrobial challenge tests. The plant manager is committed to investing in new equipment and materials that may be needed to support the project. Most of all, the manager wants to ensure that the decisions made by his company, based on data generated in the QC laboratory, use good scientific principles and state-of-the-art technologies, as expected by the regulatory agencies.

7.13
Timeline

Session 1:   Drug Development, Biofilms, Sanitization
Session 2:   cGMPs
Session 3:   Quality Control, Quality Assurance, Manufacturing Sciences, and Regulatory Groups
Session 4:   New cGMP Initiatives
Sessions 5–7:   Student Work
Session 8:   Student Presentations

7.14
Study Plan and Assignments

7.14.1
Session 1

7.14.1.1 Presentations
The instructor presents the case scenario for technology improvement and reviews the basics about biofilms and their impact in the biopharmaceutical industry.

Drug Development, Biofilms, and Sanitization (Presentation #25):

- Biofilm basics
- Overview of pharmaceutical production
- Overview of the United States Pharmacopeia
- Overview of the testing of sanitizers and disinfectants

7.14.1.2 Assignment #1
Read: Introduction to Part II

Students should review literature on topics addressed in Session 1.

The students are divided into teams for this part of the exercise, which will mostly be a homework assignment. Each team of students will be expected to research one of the following topics:

- Differences between biofilm cells and planktonic cells
- Difference between sanitization and sterilization methods for equipment decontamination
- Types of Clean-in-Place (CIP) methods used in the pharmaceutical industry for equipment cleaning (more specifically, cleaning of chromatography columns)
- Review of United States Pharmacopeia (USP) Chapter ⟨1072⟩ (www.usp.org), Disinfectants and Antiseptics*
- Review of methods for disinfectant testing described in the AOAC International, the Association of Analytical Communities (AOAC International, 1995. AOAC Official Method 960.06 Germicidal and Detergent Sanitizing Action of Disinfectants)*

7.14.2
Session 2

7.14.2.1 Presentations
The instructor reviews with the students the information from their homework assignment.

cGMPs, Qualification, and Validation (Presentation #26):

- Definition of cGMPs
- Overview of the FDA and the CFR Parts 210 and 211
- Overview of the International Conference on Harmonization (ICH) Q7, *Good Manufacturing Practice Guide for Active Pharmaceutical Ingredients*
- Overview of method qualification and validation
- Overview of equipment qualification and validation

During this lecture, the instructor may elect to create study questions designed to encourage discussion of the topics being addressed in the case, to include theory presented in prior lectures. Typically, these study questions provide supplemental material related to the topic that perhaps will not be presented in the lecture but that is critical to understanding the challenges faced by pharmaceutical/biopharmaceutical companies when implementing a new or alternate test method.

7.14.2.2 Assignment #2
Students are expected to give a brief presentation on the topics they researched and start open discussions on opportunities for improvement. They should discuss what they found out about biofilms and existing disinfectant testing as well as how these tests apply to biofilm contamination of production equipment. Students should also discuss what they found out about cleaning of chromatography columns and summarize differences between sanitization and sterilization methods for equipment decontamination.

* The instructor should make these documents available
  to students.

7.14.3
Session 3

7.14.3.1 **Presentations**
The instructor reviews with the students the progress they made from Session 2 and presents additional information that will help them with their assignment.

The Role of the Quality Unit, Manufacturing Sciences, and Regulatory (Presentation #27):

- The role of the QC, Quality Assurance (QA), Manufacturing Sciences and Cleaning Validation, and Regulatory groups in pharmaceutical manufacturing
- The role and contents of SOPs, test methods, and engineering test plans in a regulated industry

During this lecture, the instructor should point out to the students that the case being conducted impacts different departments in a pharmaceutical/bio-pharmaceutical organization. For example:

- The Manufacturing Sciences and Cleaning Validation group develops and qualifies methods for equipment cleaning.
- The QC group develops methods for testing of samples submitted by the Cleaning Validation group. QC is also responsible for developing methods for testing of sanitizers and disinfectants.
- The QA group ensures that the proposed cleaning strategies and test methods are developed and executed in accordance with cGMPs.
- The Regulatory group ensures that all functional groups adhere to the latest regulatory requirements and perform their work in full compliance with the regulations.

7.14.3.2 **Assignment #3**
Based on the information provided by the instructor, students will debate and decide which areas evaluated in their homework assignment belong in which of the four functional groups presented by the instructor. For example:

- Types of methods used for equipment cleaning (e.g., CIP methods) belong to Cleaning Validation.
- Types of methods used for testing of disinfectants/sanitizers belong to QC.
- Testing and qualification requirements defined in the CFR, ICH Q7, and the USP belong to the Regulatory group.
- Information about the proper use of sanitization vs. sterilization for equipment cleaning and the appropriateness of test methods used for qualification of sanitizers and disinfectants (e.g., biofilm vs. planktonic cells) belong to QA. This functional group also evaluates all the information presented on the equipment cleaning program to ensure testing will be executed using cGMPs.

At the end of this assignment, the instructor will divide the students into four teams that will represent the QC, QA, Cleaning Validation, and Regulatory groups. The instructor will then ensure that the students are aware of their roles as members of these four key functional teams in a pharmaceutical company. The students should focus on the following homework and classroom assignments.

- *Team 1 (Cleaning Validation)* – Students should further research best CIP methods for cleaning biopharmaceutical equipment, such as chromatography columns. This information can be obtained via the Internet, especially from Web sites for vendors of chromatography equipment and chemical sanitizers. The deliverable for this group should be a chemical cleaning method that will be used to remove biofilms from production equipment.
- *Team 2 (Quality Control)* – Students should further evaluate a method for growing biofilms in the laboratory as well as a method to test the antimicrobial properties of the chemical agents chosen by Team 1 against biofilm cells. This information can be obtained via the Internet, especially from the Web site for the Center for Biofilm Engineering (CBE) at Montana State University (www.erc.montana.edu).
- *Team 3 (Quality Assurance)* – Students should define which documents will be needed in order to execute the project. For example, to challenge the cleaning solutions against biofilm cells, an engineering test plan or a qualification protocol should be created; if new equipment will be used, an equipment SOP should be created; a test method will be needed for growing biofilms in the laboratory. These students should also recommend which test organisms should be used in the study in order to reflect typical equipment contaminants or typical flora in a pharmaceutical environment.
- *Team 4 (Regulatory)* – Students will evaluate CFR Parts 210 and 211 and ICH Q7 documents to identify which sections apply to the work that will be performed to implement an alternate and improved technology in the laboratory, a new piece of equipment, or an improved equipment cleaning method.

7.14.4
Session 4

7.14.4.1 **Presentations**
The instructor will discuss the progress of the assignments and present his or her expectation for an interim team report that will include some preliminary identified technology or method improvement opportunities. Reports should include approximate cost for any new piece of equipment and materials that need to be purchased. The instructor will also present an overview of the FDA initiative for cGMPs for the 21st century.

FDA Initiatives for cGMPs in the 21st Century (Presentation #28)

7.14.4.2 Assignment #4

Students are expected to continue working on their assignments and start making decisions for their recommendations as defined in their project deliverables. The main deliverable at this stage of the project is an interim report that should be due the following class. This report is a critical deliverable in case of unexpected situations that will prevent the project from being completed in the agreed timeframe or in case an alternate path must be taken. Interim reports must include any costs that the company would have to incur to implement the new technology or methodology. This information should be provided in the form of a cost vs. benefit analysis.

7.14.5
Session 5

7.14.5.1 Presentations

The instructor will review with the students their interim reports. Playing the role of plant manager, the instructor will decide whether the proposals and findings are suitable for the project and are solid in terms of withstanding regulatory scrutiny during an inspection. The instructor will decide whether the proposals are a "Go" or "No Go" and whether he or she agrees with the findings from the quality and regulatory groups.

7.14.5.2 Assignment #5

The students will evaluate the feedback from the instructor and start refining their proposals and summary of findings. The following are examples of deliverables at this stage of the project:

- *Team 1 (Cleaning Validation)* – Students will work on a proposal for a new cleaning method for chromatography columns that have become contaminated with biofilms.
- *Team 2 (Quality Control)* – Students will work on a proposal for a new method to test sanitizers using biofilms grown in a laboratory.
- *Team 3 (Quality Assurance)* – Students will prepare a list of documents that will be required to implement the improved technology or methodology.
- *Team 4 (Regulatory)* – Students will prepare a document that states which sections of the CFR Parts 210 and 211 and/or ICH Q7 guide apply to the project designed to implement an improved technology or manufacturing equipment cleaning methodology.

7.14.6
Session 6

7.14.6.1 **Presentations**
The instructor provides students with any additional support needed to ensure completion of their deliverables in the expected timeframe.

7.14.6.2 **Assignment #6**
Students continue to work on their assignments and start preparing the final group report.

7.14.7
Session 7

7.14.7.1 **Presentations**
The instructor reviews the documentation prepared by the students and provides feedback. The instructor should also provide clear expectations on what is required for the final team reports that will summarize the various assignments conducted by the four teams of students.

7.14.7.2 **Assignment #7**
Students finalize their deliverables and work on their final team report and a PowerPoint (slide show) presentation for the last day of class.

7.14.8
Session 8

Instructor Tasks: This is the last day of class, when the students present their work to the instructor and their classmates. The instructor must ensure provisions are made ahead of time for this event. Just as in Part I, some of the recommended tasks for the instructor are:

- Coordination of logistics for the presentations
- Invitation of guests to the event
- Preparation of questions for the groups to test their knowledge of the subjects covered during the course of the project
- Evaluation of the groups' presentations based on content, innovation, and skills
- Ensuring that questions are posed to the groups to test their ability to answer questions from an audience

7.14.8.1 **Assignment #8**

This is the day when the project ends and the students will be evaluated based on their deliverables:

- Delivery of all the work collected and prepared, which must include a written final report and proposal for improvement
- Summary of work performed in the form of a PowerPoint presentation
- Preparations in anticipation of questions posed by audience and instructor

7.15
Company Supervised Practicum

This case can be conducted in a classroom setting as detailed in this chapter or as a supervised practicum in a pharmaceutical/biopharmaceutical company, where students may actually work in the laboratory to implement a new technology. In this case, the project might be delayed by two to three weeks due to needed preparations and training for the students as discussed earlier in this chapter. In addition, the project may be delayed even further if new equipment must be purchased and funds are not immediately available. Microbiological testing is also time-consuming and often generates unexpected results. If students need time to investigate an alternative microbiological technology, the project may take longer than the time allotted for the case. Therefore, the best approach is to either divide the case into small projects assigned to separate teams of students, prioritize the project deliverables, or carry out the project in successive block periods during two school semesters. In some cases, it may be necessary to redefine deliverables as the project progresses throughout the semester.

A similar case to the one described above was carried out at a biopharmaceutical manufacturing company. The director of microbiology wanted to implement a new technology for growing biofilms in the laboratory and later use the new method to rechallenge the chemical sanitizers they used to clean production equipment.

The students used a method developed for *Pseudomonas aeruginosa* biofilms using a biofilm reactor[5] as a model for their project. Typically, a biofilm is grown in a well-controlled biofilm reactor that contains coupons representing various types of equipment surface materials. After the biofilm is grown on the coupon surface (typically 48 hours), the coupons are exposed to the antimicrobial solution for a defined period of time. Exposure can be static or under flow conditions. Following exposure to the biocidal agent, the coupons are rinsed with a neutralizing solution, and the biofilm is removed and transferred to a

5) ASTM International. "Standard test method for quantification of a *Pseudomonas aeruginosa* biofilm grown with high shear and continuous flow using CDC biofilm reactor, E-2562-07". *ASTM Book of Standards*, vol. 11.05. ASTM International, West Conshohocken, PA, 2007.

sterile container. The sample preparation is then homogenized and serial diluted for plating and enumeration of viable cells using optimized microbiological test methods. Disinfectant efficacy is determined by comparing the number of viable cells remaining on a disinfectant-treated coupon to the number of viable cells on an untreated coupon (control).

Since the test protocols require that the biofilm formed on the coupon surface be subjected to various types of manipulations to ensure neutralization of the disinfectant and removal of biofilm mass from the test surfaces, method qualification and optimization for the various types of test organisms are required. The equipment chosen was the CDC (Center for Disease and Control) biofilm reactor, designed by the CDC and manufactured by BioSurface Technologies Corporation (Bozeman, Montana; www.biofilms.biz). A thorough assessment and understanding of the relative contributions of the kill and removal steps, to include potential for cell wash-off during the neutralization step, must be performed. The observed number of cells lost during wash-off and recovery procedures can be used to calculate the variability in test results. Indeed, much time can be spent in the laboratory attempting to qualify this type of microbiological method.

In this situation, the company had already selected the technology and piece of equipment to be implemented. However, the actual method needed to be adapted and optimized to meet the company's needs.

The project was assigned to a team of three microbiology students, who were supervised by a company employee, the manager of the QC microbiology laboratory. Figure 7.5 contains the abbreviated project plan created by the QC manager. This document formed the basis for the work that was performed by the students.

The students who participated in this project were required to undergo a total of 10 hours of training in cGMPs, company safety (to include laboratory safety procedures), and documentation practices. Students spent at least three hours a week at the sponsor site. However, they were expected to dedicate about 10 hours a week to the project. The time outside the company was used for the review and creation of documents, meetings with microbiology professors, and analysis of data. On site, students spent most of their time assembling the biofilm reactor, contacting company subject matter experts (e.g., cleaning validation and manufacturing personnel) and vendors, and performing laboratory testing for optimization of the test method.

On some occasions, students had to modify their class schedule to be able to conduct the microbiological experiments that were expected to last several hours. On one occasion, students worked in shifts (i.e., one student came in the morning, another in the afternoon, and the other in the evening). For this project, the company project leader had to request assistance of other laboratory personnel to make sure that when students arrived on site for their scheduled class, the microbiological media had been prepared and test organisms had been subcultured. A laboratory-based case, especially one that involves microbiological testing, is indeed very time-consuming, not only for the students but

Biofilm Project Plan
Prepared by Joan Doe
Manager, QC Laboratory

Purpose: The purpose of this project is to prepare the documentation needed to test sanitizers against biofilm organisms that are typically isolated from production equipment and manufacturing environment.

Goal: The goal of this project is to adapt a biofilm reactor to best mimic equipment sanitization processes used in manufacturing and to optimize the procedure for growing biofilms for typical company environmental isolates.

Scope: The scope of this project is to prepare the needed documentation to have a program to test sanitizers against biofilm cells. However, if time permits and equipment is available, challenge studies for the various sanitizers may also be performed.

Deliverables:

1. *Equipment SOP*, in draft, for the chosen biofilm reactor
2. *Test Method*, in draft, for growing biofilms in the laboratory; optimize procedure for gram-positive cocci, gram-negative rods, and spore-forming bacteria (consult with laboratory management for test organisms to be used in the study)
3. *Engineering Test Plan (ETP),* in draft, for evaluation of antimicrobial effectiveness of various sanitizer solutions against biofilm organisms

Milestones:

- Week 1: Project begins
- Week 3: Complete draft SOP for the biofilm reactor
- Week 5: Complete draft ETP for evaluation of antimicrobial effectiveness of sanitizers
- Week 7: Complete draft Test Method for growing biofilms in the laboratory
- Week 9: Complete method optimization for environmental isolates
- Week 10: Make presentation to company management and complete project deliverables

Fig. 7.5 Example of an abbreviated project plan for a science-based case.

also for the instructors and project leaders. Therefore, careful planning and se-lection of achievable milestones and project deliverables must be in place for this type of case to be successful.

The project at the sponsor company was completed within the 10-week peri-od, and all deliverables were provided to the company project leader as planned. There were some delays due to the late arrival of the purchased equipment. Also, as expected, delays were encountered during method optimization for the company environmental isolates, and some of the testing had to be repeated more than once. Despite the expected and unexpected delays, the students were able to assemble the biofilm reactor and complete the draft documents as speci-fied in the project plan. Figure 7.6 is a schematic representation of the set up for the biofilm reactor assembled by the students.

According to some of the students, testing delays and repeat work were chal-lenges. Students said they would have liked to have had the time to actually per-form the sanitizer efficacy studies using the biofilm reactor. On the positive side, all students enjoyed the dynamics and fast-paced nature of a laboratory-based case and the opportunity to perform work in a QC laboratory at a bio-pharmaceutical company. The students stated that factors contributing to the completion of the project in the timeframe specified in the project plan were good planning, preparation, multitasking, and flexibility.

This case also confirms the benefits not only to a sponsor company but to the real-life learning experience provided to students. The idea of using laboratory-based cases provides the opportunity for a dynamic style of teaching, while stu-dents are required to demonstrate team work behavior in order to complete the assigned tasks. This is an excellent way to incorporate soft skills and behavior-based teaching.

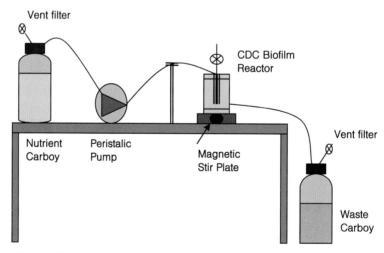

Fig. 7.6 Schematic of a typical set-up for the CDC biofilm reactor.

7.16
Conclusion

In a highly regulated environment, compliance is always at the forefront. Therefore, in order to attain excellence in operations, a pharmaceutical/biopharmaceutical company must find novel ways to integrate compliance into its manufacturing infrastructure rather than viewing compliance as a separate process. One approach is to create business platforms that combine both compliance and continuous improvement initiatives. In fact, one of the hallmarks of a company that has become truly "lean" is a focus on continuous improvement. Another approach is to apply risk management to help the company operate at optimal efficiency and produce high-quality products and test results at the lowest cost. This strategy can be used to optimize quality control in the manufacturing process to improve production rates and test methods and to reduce cycle times. Continuous improvement, whether of processes or methods, is in fact one of seven key principles in lean manufacturing and one that must be made visible in any organization.

8
Aligning Behaviors and Standards in a Regulated Industry: Design and Implementation of a Job Observation Program

Amy Peterson and John Shaeffer

Contents

Industry Immersion Learning. Real-Life Industry Case-Studies in Biotechnology and Business
L. Borbye, M. Stocum, A. Woodall, C. Pearce, E. Sale, W. Barrett, L. Clontz, A. Peterson, J. Shaeffer
Copyright © 2009 WILEY-VCH Verlag GmbH & Co. KGaA, Weinheim
ISBN: 978-3-527-32408-8

8.1
Mission

The mission of this case study is to demonstrate how to manage human behavior variability within the manufacturing environment using a behavior modification program called a Job Observation Program.

8.2
Goals

The goals are for the students to design and implement a behavior modification program, including the following components:

- Identification of misalignments between employee behaviors and company-defined rules (standards)
- Assessment of whether the company-defined rules (standards) or the employee behaviors need modification
- Design and implementation of a positive reinforcement plan to promote desired behavior modification
- Feedback monitoring of the effects of the program implementation

8.3
Predicted Learning Outcomes

Students will:

- learn about the concept, application, and impact of a job observation program;
- attempt to create consistency between selected personnel and their work by modifying and controlling their behaviors and creating procedural adherence.

8.4
Introduction

"We are what we repeatedly do. Excellence, therefore, is not an act, but a habit."

Aristotle

8.4.1
Human Error and Human Error Prevention by Job Observation

Human fallibility is an inherent attribute to any work environment and it must be properly managed. Preventing human errors from resulting in catastrophes is important to everyone because inevitably one could be the customer with spoiled food, patient with a botched surgery, or passenger nervously awaiting a smooth airplane trip to visit family.

In the pharmaceutical industry there are processes, which are both prone to human error and unforgiving of human error, i.e., if a mistake is made there are very large consequences. For example errors made during manual operations to produce a batch of medicine that result in the loss of a product batch have both consequences that are financial and humanitarian; the company loses product and money, and the patient does not receive medication.

Preventing human error from negatively impacting the manufacturing process is essential for a pharmaceutical company. A tremendous amount of effort is directed towards controlling human error to prevent deviations. Company policies and procedures are developed to help employees achieve consistency and reduce errors. The primary way the pharmaceutical industry ensures compliance and reflects on errors is by review (after the fact).

Job Observation Programs use positive, real-time reinforcement to modify employee behaviors such that they become low risk behaviors and to catch errors before they lead to deviations. Good programs can be designed based on the knowledge of certain behaviors: Behaviors can predict an increase in the risk that an error will occur. Many automobile insurance companies use this approach. When a driver receives multiple speeding tickets the company will typically increase the insurance rates. This is due to the fact that speeding is a behavior that increases a chance for an accident. The driver may not yet been involved in an accident but the company will increase their rates due to increased risk based on the number of tickets received.

8.4.2
Procedural Adherence and Human Behavior

Regulatory bodies define rules (regulations) to which facilities and/or manufactures must adhere. One significant rule usually defined is to "follow established procedures" (instructions or rules). Procedures are contained within controlled documents that provide the necessary instructions to ensure that multiple humans perform a task in the same consistent manner. Adherence to established procedures in a regulated environment is essential and fundamental to maintaining regulatory compliance and quality control of a manufacturing process. Regulated industries spend significant time and money investigating the reasons procedures are not followed.

Procedural adherence is a topic that the Federal Aviation Administration (FAA), National Transportation Safety Board (NTSB), and the Nuclear Regulatory Commission (NRC) enforce with great vigor. The NRC has managed procedural adherence for commercial nuclear operations since the mid 1980s, when the NRC collaborated with experts in various disciplines, such as psychology, cognitive engineering, industrial engineering, and management to further understand how to manage human fallibility and variability related to following established procedures, ensuring procedural adherence. These industries have effectively applied the necessary structure and concepts to the work environment and documents to achieve success.

Currently the FDA's published trends show that one of the most prevalent challenges throughout the pharmaceutical industry is lack of procedural adherence. Ensuring that the hundreds of thousands of pharmaceutical/biotechnology employees follow established procedures that describe their controlled, validated processes to manufacture medications is a crucial element to ensuring the purity, safety, and efficacy of medications for the population. One may speculate that the FDA will develop similar regulatory expectations with respect to mitigating detrimental effects due to human error. Controlling the human element of a manufacturing process poses difficult challenges.

8.4.3
The Necessity for Job Observation

The previous case study (Chapter 7) described Operational Excellence and various ways to improve processes and techniques. These improvement efforts usually require change in employee behaviors. Often, the employees receive such training and are expected to utilize these behaviors immediately. However, experience has shown that if there is not a formalized program implemented within the work area which will embed the behavior changes necessary the employees often revert back to their previous behaviors. Achieving the proper behavior change is critical for error prevention.

The consequences for deviations between stated compliance and actual actions can manifest into significant variations of the manufactured product. Deviations can result in audit findings, warning letters, consent decrees, and other enforcement actions. In extreme cases it can lead to plant closure and bankruptcy. Job observation programs can help minimize the risk of deviations by immediately interjecting an intervention strategy before significant deviation can occur. It is essential that the employee is positively rewarded for making the appropriate adjustments and the program does not have a negative impact for the employee. The employee will welcome the observer when positively rewarded and the program will observe true behaviors instead of adjustments only when an observer is present. Negative observation programs usually fail because the organization does not observe true behaviors.

8.5
Case Scenario

Pharmaceutical and biotech manufacturing facilities are found around the world. Managers of these facilities typically possess a technical or scientific background that prepares them to address and correct problems within the manufacturing process that related to the technical or scientific aspects of the process. As yet, there are limited resources available to managers to help control human fallibility and variability. Manufacturing steps often require a second individual to ensure a step was performed and performed correctly, documenting these activities on the record as a "checker" or "checked by" as defined by the company.

The process to develop vaccines utilizes a number of manufacturing steps. These steps include (but are not limited to) fermentation, conjugation, purification, and filtration. Manufacturing steps require that technicians perform manual manipulations of various components to successfully complete the processing.

The manufacturing process steps are written in a controlled document, which describes the steps that must be performed to manufacture a vaccine. The steps to manufacture a vaccine must be in accordance with the FDA License to Manufacture and therefore must be followed to maintain regulatory compliance as stated in the Code of Federal Regulations (CFR). The controlled document that contains the written steps to manufacture a vaccine is typically referred to as a batch record.

Some of the components technicians manipulate include tanks, lines, valves, and clamps. The tanks are installed in an area either permanently or temporarily. Tanks that are temporarily installed are on wheels and can be moved from location to location. The physical lines that the liquid or steam travels through are either permanent or temporary. The permanent lines are sometimes referred to as hard-piped lines, while the temporary lines are sometimes referred to as flexible hoses. The valves are either permanently attached or are temporary and have to be manually attached to the lines. Clamps are used any time the component (tank, line, or valve) is manually attached.

It is important that technicians utilize clean components and aseptic techniques in the manufacturing process. The permanent components are cleaned by initiating a flow of high-temperature steam through permanent lines to the component. The temporary components are cleaned either by initiating a flow of high-temperature steam through permanent lines to the component at predefined locations or by placing the component in a washer and then exposing it to autoclaving. Usually the larger components, such as temporary tanks and flexible hoses, are cleaned using the permanent lines, and the smaller components, such as the valves and clamps, are placed in the washer and autoclave.

There are several steps in this entire process, which require control. Examples as required by the CFR are in sections 211.103 Calculation of yield and 211.101 Charge-in of components.

1. The batch shall be formulated with the intent to provide not less than 100 percent of the labeled or established amount of active ingredient.
2. Components for drug product manufacturing shall be weighed, measured, or subdivided as appropriate. If a component is removed from the original container to another, the new container shall be identified with the following information:

   - Component name or item code;
   - Receiving or control number;
   - Weight or measure in new container;
   - Batch for which component was dispensed, including its product name, strength, and lot number.

3. Weighing, measuring, or subdividing operations for components shall be adequately supervised. Each container of component dispensed to manufacturing shall be examined by a second person to assure that:

   - The component was released by the quality control unit;
   - The weight or measure is correct as stated in the batch production records;
   - The containers are properly identified.
   - Each component shall be added to the batch by one person and verified by a second person.

How personnel interact between actual procedure step performance and required documentation of procedure step performance and a check or verification within a manufacturing process creates a complex system in which controlling the human interaction presents a unique challenge. The same techniques used in a job observation program to control human fallibility and variability in manufacturing a drug and/or a vaccine are applicable to other processes within many different industries, such as healthcare, transportation, and, as previously discussed, aviation and nuclear energy.

This case is designed to demonstrate management of personnel behavior within a less automated, more manual manufacturing process, one where the human element is a substantial variable to control. General procedure usage and its associated behaviors were chosen.

8.6
Timeline

Session 1–5:   Job Observation Program Development
Session 6:   Implementation of Job Observation Program
Session 7–8:   Evaluation, Modification, and Presentation
               of Job Observation Program

8.7
Study Plan and Assignments

8.7.1
Session 1

8.7.1.1 Assignment #1
Read: Introduction to this chapter

   Identify "programs" and behaviors that minimize human error (examples: getting the right orders to the right customers at a restaurant; getting the correct medicine at the pharmacy, etc.) (Reference section 8.4).

8.7.2
Session 2

8.7.2.1 Presentations
Review and discuss job observation programs from Assignment #1 (Reference section 8.4)

What does the program do?
How does the program work?
Why do we need a job observation program?

8.7.2.2 Assignment #2
Extrapolate findings from Assignment #1 to a manufacturing process. Identify behaviors to monitor.

8.7.3
Session 3

8.7.3.1 **Assignment #3**
1. Determine and select the area (procedure usage, procedure adherence, procedure documentation, etc.) for which a list of specific desired behaviors will be developed.
2. Determine the goal or result that will be achieved by moderating existing behaviors or developing new behaviors. Why will monitoring and "dispositioning" (modifying) behaviors accomplish the goals?
3. Develop a list of the specific desired behaviors to reinforce in the work place and determine the setting where these behaviors will be emphasized (manufacturing floor, laboratory, office, etc.).
4. Define standards associated with tasks within the area of monitoring.

8.7.4
Session 4

8.7.4.1 **Assignment #4 (Reference section 8.8, Appendix A and B)**
Develop a job observation program to monitor specific desired behaviors and a method to determine and correct behaviors not in alignment with desired behaviors.

8.7.5
Session 5

8.7.5.1 **Assignment #5 (Reference section 8.8)**
1. Develop metrics to determine the effectiveness of a job observation program.
2. Compare program to the items needed for a widely applicable job observation program. Does it meet the criteria?

8.7.6
Session 6

8.7.6.1 **Assignment #6 (Reference section 8.8)**
1. Implement program.
2. Test program

8.7.7
Session 7

8.7.7.1 **Assignment #7 (Reference section 8.8)**
1. Evaluate program (see "Job Observation Program Evaluation" Section 8.9).
2. Develop modification plan based on evaluation.
3. Answer questions in Appendix C.

8.7.8
Session 8

8.7.8.1 **Presentations**
Presentation of job observation program, implementation, evaluation, and modification.

8.8
Items Needed for a Widely Applicable Job Observation Program

1. *Effective expectations* should include
   - identification of all individuals required to apply the expectations;
   - actions to be carried out and standards for performance;
   - cues that signal action;
   - reinforcement plan to motivate future application of expectation.
2. *Actions* should
   - satisfy a specific definition (what is acceptable and what is unacceptable?);
   - be objective and concise, not open to interpretation;
   - be observable, countable;
   - be doable, possible to perform in the work environment.
3. *Job observers* should
   - be sensitive to job site conditions as well as individual behaviors;
   - optimize the effectiveness of data gathered; be perceived as resources rather than threats.
4. *Job observation/inspection activities* (points to monitor) may include
   - checking work against the work plan;
   - assessing presence of obstacles to performance;
   - determining if scope of work has changed;
   - verifying availability of appropriate tools;
   - monitoring overall work environment;
   - verifying if workers accurately perceive risk and priorities;
   - reinforcing and coaching workers about observed behaviors.
5. A *reinforcement plan* determines
   - desired results for selected task(s);
   - specific behaviors to obtain results;
   - measures of results in terms of quality, quantity, costs, or timeliness;
   - feedback about past behavior that can help an individual change (feedback should be specific, sincere, immediate, and personal);
   - positive reinforcement plan for pinpointed results and behaviors.
6. *Immediate feedback* is desired through a defined feedback mechanism, which must be
   - easy to monitor and maintain;
   - specific.

7. *Implementation plan* should define timing and the implementation method.
8. *Metrics* to determine effectiveness must be objective and measurable.

8.9
Job Observation Program Evaluation

Each of the following attributes of the job observation program should be addressed:

1. Setting where the observations take place (manufacturing floor, on the job, in a simulated setting, etc.)
2. Specific behaviors to observe (what work behaviors – safety, work site behaviors – is the program trying to modify and how will modification be accomplished?)
3. Development of the documentation process to use to track the observations and effectiveness of the program (filling out a paper form, electronic form, none but the number of observations, etc.)
4. The feedback mechanism your program will use, including the timing of the feedback (on-the-spot verbal corrections, written reports given at monthly group meeting, etc.)
5. Specific criteria used during the observations (will different observers come to the same conclusions?)
6. The reward or punitive system to implement with the program (money, free food, verbal recognition, negative documentation in personnel files including reprimand for a specified number of negative comments)
7. Method to measure program effectiveness, to include metrics and goals (will you recognize only absolute success or are there grades of success associated with the program?)
8. Implementation plan for the program
9. Duration and evaluation of the program
10. Changes to the program based on evaluation

Appendix A:
Job Observation Program for a Commercial Kitchen

The production of a pharmaceutical drug can be characterized as following a recipe, a procedure. Cooking is, therefore, used as an example to further understand the concept of observing behaviors. The behavior standard, the metrics, an example of a cooking observation form, and an elaboration of items needed for a cooking observation program are given below.

A1
Behavior Standards

1. Always follow the recipe and perform the tasks in the order specified (100% of the time).
2. The utensils used should always be clean and never reused for a different recipe unless first washed in the dishwasher.
3. When specific amounts are required by the recipe, ingredients are always measured in approved measuring devices.
4. Approved timers are always used and set to within ±5% of specified times.
5. Food is checked during cooking within ±5% of specified cooking times to ensure it meets recipe requirements.

Observations will be recorded and the results of the individual observations will be given to the corporate manager. The manager will enter data into an MS Excel spreadsheet and generate monthly reports. Written comments will be compiled, analyzed for trends, and included on the monthly report.

A2
Metrics

1. Maintain above 95% satisfaction on any food inspections (monthly corporate inspections; annual state regulatory inspections)
2. Maintain customer satisfaction above 98% (2 complaints per 100 people)

Cooks who achieve the metrics for three consecutive months will be given a bonus of $ 300.

A3
Example Form: Job Observation – Commercial Cooking

Example Form: Job Observation – Commercial Cooking

Recipe Title _____

Were all recipe steps completed?   ❑ Yes   ❑ No
   In no, list which steps were missed:_____

Were clean utensils used? ❑ Yes   ❑ No
   If no, list all utensils used:

_____

_____

Were ingredients measured according to specifications?   ❑ Yes   ❑ No
   If not, list ingredient and alternate method used to measure ingredients:

_____

_____

_____

Were all steps performed in order when prompted?   ❑ Yes   ❑ No
   If no, list which steps were performed out of sequence:

_____

_____

_____

Was the food inspected at the proper cooking time?   ❑ Yes   ❑ No
   If not, list the proper time and the actual time the food was inspected.
             Proper time _____
             Time of inspection: _____

Was the timer set to within +/- 5% of cooking time?   ❑ Yes   ❑ No
   If no, list the steps when timer usage was missed.

_____

_____

A4
Items for the Basis of a Cooking Observation Program

1. *Effective expectations* contain the following:
   - Specific group of people to whom expectations apply: Cooks in a commercial kitchen
   - Actions to be carried out and standards for performance:
     – Actions: Cooking recipe in the kitchen
     – Standards for following recipe: Cooking times, usage of utensils, measuring devises, and order of tasks
   - Cues that signal action:
     – (Implied:) Cooking starts when an order comes in
        Note: Could go back and modify program to include this
   - Reinforcement plan to motivate future application of expectation:
     – Bonus to cooks who achieve the metrics for three consecutive months.
2. *Actions* should
   - satisfy a specific definition (what is acceptable and what is unacceptable?)
   - be objective and concise, not open to interpretation
   - be observable, countable
   - be doable, possible to perform in the work environment

The cooking observation program meets all of the above criteria ["Always follow the recipe and in the proper order when specified (100% of time); utensils used are always clean, never reused in a different recipe unless washed in the dishwasher; ingredients are always measured when specific amounts are required by recipe; food is checked during cooking within ±5% of specified times, and timers are always used and set to within ±5% of specified times"].

3. *Job observers* should
   - be sensitive to job site conditions as well as individual behaviors:
     Observers should maintain 5 feet distance from the cook to avoid interference. They should not provide feedback until the end of the day unless feedback is critical for meeting metrics. Examples are overcooking of the food, using the wrong ingredients, etc.

   - optimize the effectiveness of data gathered, be perceived as resources rather than threats:
     If observers provide feedback to the cooks such as that the cooks are running over on their cooking time or a wrong ingredient is about to be used, would that help or hurt the cooks in achieving their metrics? If the observation feedback helps the cook achieve the bonus, then the observer will be seen as a resource. If the observer makes observations that prevent the cook from getting the bonus, then the observer will be seen as a threat.

4. *Job observation/inspection activities* (points to monitor) may include
- checking work against the work plan;
- assessing the presence of obstacles to performance;
- determining if the scope of the work has changed;
- verifying the availability of appropriate tools;
- monitoring the overall work environment;
- verifying if workers accurately perceive risk and priorities;
- reinforcing and coaching workers about observed behaviors.

If an observer sees a cook reusing a utensils for an unrelated recipe without washing it first, the observer takes note of this event as an error. At the end of the day, the observer may chose to discuss the observation with the cook. The cook may have an explanation, such as that he or she was out of clean utensils and did not want to have the recipe ruined waiting for the dishwasher. Therefore, the cook made a conscious decision to utilize utensils that were not washed in the dishwasher. This behavior would still count as a procedural adherence deviation, a negative behavior observation (not meeting standards). However, because the employee was not given the proper tools, it may be that corporate management is responsible for creating this condition (not enough tools). Since this cook had to perform a "workaround" to get the job done (and the cook got the work done), reusing the utensils would be a negative behavior seen in the light of the observation program but most likely not impact the cook's goal of achieving the bonus because the necessary tools were not in place.

Note: With respect to the procedural adherence case, if employees are not following procedures 50% of the time, and it is observed that 90% of the time when they do not follow the procedure it is because it is not possible to follow the procedure, this still shows up as negative behavior but not one that the employees can change (other than the 10% due to the employees).The organization would have the responsibility to alter the procedure.

5. A *reinforcement plan* determines
- desired results for selected task(s);
- specific behaviors to obtain results;
- measures of results in terms of quality, quantity, costs, or timeliness;
- feedback about past behavior that can help an individual change (feedback should be specific, sincere, immediate, and personal);
- positive reinforcement plan for pinpointed results and behaviors.

The behaviors monitored will achieve the desired results provided that corporate management provides good recipes, necessary equipment, and proper resources. If desired results (the metrics) are not obtained, the organization can easily deduce whether the cook (low observations rating) or the organization (high observations rating but desired results not obtained) is responsible. See response to point 8 below.

6. *Immediate feedback* is desired through a defined feedback mechanism, which must be:
   - easy to monitor and maintain;
   - specific.

The program is easy to administer and feedback is immediate if it has an impact on the metrics or, if there is no immediate impact, slightly delayed in order to avoid interfering with production. The feedback form is contained within one page and is easy to use.

7. *Implementation plan* should define timing and the implementation method.
   As an example, cooks could be observed once per week during a specified period and during the process of using all of the recipes specified. The observer would fill out the cooking observation form.

8. *Metrics* to determine the effectiveness of the program:
   - Ensure above 95% satisfaction on any food inspections.
   - Ensure clean utensils are used and maintained available.
   - Maintain customer satisfaction above 98%.

Following recipe (procedures) and ensuring the recipe measurements are used, cooking times maintained, utensils immediately cleaned, and recipe steps followed in the proper order should ensure both food inspection satisfaction and customer satisfaction.

If the program is implemented and the observer finds, for example, that the cooks are not meeting their goals due to low customer satisfaction, an investigation must take place. One or more of the following items would need correction.

- Behaviors: aberrant behaviors prevented the metrics from being met.
- Recipes/work environment: either the recipes or the work environment prevented metrics from being met.
- Job observation program: metrics are not being met but behavioral standards are being met, and the corporation is supplying all of the tools for success (good recipes, proper equipment, etc.). Consider modifying behavioral standards.

Example: Given the hypothetical scenario that only the food produced using recipe 3 turns out to receive all of the complaints, an analysis of the data from the observation program should help to identify the root cause of the problem. If during the time the complaints occurred the cooks were meeting the expectations of the program (their metrics), it would suggest that the recipe is the root cause and not the cooks. Without the job observation program, it would be difficult to decide which was the root cause: the recipe authors or the cooks. The authors would most likely defend their recipes and imply that the cooks were the root cause of the problem. The cooks would most likely defend their cooking and imply it was the recipe.

The following two items are not specifically addressed in this example:

9. *Duration and evaluation* of the cooking observation program
10. *Changes* to the cooking observation program based on evaluation

Appendix B:
Job Observation Program for GMP Documentation in a Manufacturing Facility

This example will focus on behaviors associated with GMP documentation during a manufacturing process. The behavior standards, metrics, an example of a GMP documentation observation form, and an elaboration of the items needed for a GMP documentation observation program are described below.

B1
Behavior Standards

1. Always follow GMP documentation standards as defined in the standard operating procedure (SOP) (100% of the time).
2. If a manufacturing step and its associated documentation cannot be performed by the standards defined in the SOP, then stop and notify a supervisor.

Observations will be recorded and the results of the individual observations will be given to each shift supervisor. The shift supervisor or designee will enter the data into an MS Excel spreadsheet and generate monthly reports. Written comments will be compiled, analyzed for trends, and included on the monthly report.

B2
Metrics

1. Stretch goal: Achieve a 10% reduction in documentation errors per million opportunities per month
2. Achieve fewer than 100 documentation errors per million opportunities per month

For any shift that achieves metric 2, each person on the shift will receive a $ 50 gift certificate. For any shift that achieves metric 2 for at least 6 months, each person on the shift will receive a $ 100 gift certificate. For any shift that achieves metric 1, each person on the shift will receive a $ 300 gift certificate and an Excellence Award from the plant manager.

B3
Example Form: Job Observation for GMP Documentation in a Manufacturing Facility

Example Form:
Job Observation – GMP Documentation in a Manufacturing Facility

Monitored by: _____

Date:_____ Start time: _____ Stop Time: _____

Shift Observed: 1st_____ 2nd_____ 3rd _____

Manufacturing Process Observed: _____
_____
_____

Manufacturing Batch Record in Use:: _____
_____
_____

Were the Documentation Standard Followed?: _____
_____
_____

Was a Supervisor Notified if the Documentation Standard Could Not Be Followed?:_____
_____
_____

B4
Comparison: Manufacturing Observation Program vs. Standards

B4.1   Items for the Basis of an Observation Program
1. *Effective expectations* contain the following.
- Specific group of people required to whom expectations apply: Manufacturing technicians
- Actions to be carried out and standards for performance:
  Actions: Document completion of manufacturing steps
  Standards: Defined by company-specific procedures (example: complete documentation of a manufacturing step prior to initiating the next processing step, unless specified otherwise in a manufacturing batch record)

This example has two behavior standards:

1. Always follow GMP documentation standards as defined in the standard operating procedure (SOP) (100% of the time).
2. If a manufacturing step and its associated documentation cannot be performed by the standards defined in the SOP, then stop and notify a supervisor.

- Cues that signal action:
  When a manufacturing batch record is initiated
- Reinforcement plan to motivate future application of expectation:
  1. Stretch goal: Achieve a 10% reduction in documentation errors per million opportunities per month
  2. Achieve fewer than 100 documentation errors per million opportunities per month

Positive rewards: For any shift that achieves metric 2, each person on the shift will receive a $ 50 gift certificate. For any shift that achieves metric 2 for more than 6 months, each person on the shift will receive a $ 100 gift certificate. For any shift that achieves metric 1, each person on the shift will receive a $ 300 gift certificate and an Excellence Award from the plant manager.

2. *Actions* should satisfy the following criteria:
- Specific – concise definition: what is acceptable and unacceptable
- Objective – not open to interpretation
- Observable – countable
- Doable – ability to perform in the work environment

The job observation program meets all of the above criteria ["Always follow standards (100% time) or contact supervision when the standards cannot be implemented as stated; the standards are objective since they are applied 100% of the time or supervision is notified prior to a deviation; the standards can be easily monitored using an observer by comparing the actions of the manufacturing technician to the batch or compound record; and the documentation must be completed, but there is an alternate path for the manufacturing technician if this cannot be accomplished"].

3. *Job observers* should:
- Be sensitive to job site conditions as well as individual behaviors: Although not specifically stated in this document, there would be rules for the observer to follow during observation, such as:
  - Observers should maintain 5 feet from the manufacturing technicians during task performance to avoid interference.
  - Observers should not provide feedback until the end of a shift unless feedback is necessary to prevent a manufacturing batch from being lost; for example, an individual is about to add the wrong ingredient.
- Be perceived as resources rather than threats, thus optimizing the effectiveness of data gathering.

If observers provide feedback to the manufacturing technicians, such as "You missed a signature," or "You performed steps out of sequence," will the observation help or hinder the manufacturing technicians in achieving their metrics? If the observation feedback helps technicians do the job correctly and therefore receive the positive reinforcement (in this example, a gift certificate), then observers will be seen as a resource and their help desired. If the observers make observations that prevent technicians from receiving positive reinforcement (gift certificate), then the observers will be seen as a threat and will not be well received. In this example, catching a documentation error during processing will prevent the error from being identified later in the final quality assurance review process, at which time the error will have a negative impact on investigations and processing time for releasing batches for commercial usage. This allows the observers to be seen as a resource to manufacturing technicians as well as the company as a whole.

4. *Job observation/inspection activities* (points to monitor) may include
   - checking work against the work plan;
   - assessing the presence of obstacles to performance;
   - determining if the scope of the work has changed;
   - verifying the availability of appropriate tools;
   - monitoring the overall work environment;
   - verifying if workers accurately perceive risk and priorities;
   - reinforcing and coaching workers about observed behaviors.

If an observer sees manufacturing steps being documented out of sequence in relationship to task performance, and the record does not allow the documentation to be performed out of sequence, then these discrepancies must be evaluated and addressed. At the end of the shift, observers will discuss the outcome of their observations with the manufacturing technicians. Let us consider one possible response for the observation discrepancy. The manufacturing technicians may explain that they had to perform the documentation out of sequence with task performance at that time because if they had taken time to contact the supervisor, the batch would have been lost due to processing time constraints. Therefore, the technician made a decision not to meet the documentation standards. Because the manufacturing technician had to perform a workaround to get the job done, it would still be a "negative observation data point" for the observation program, but the organization would take responsibility for correcting the discrepancy. The negative observation would not impact the manufacturing technician's performance goals for obtaining a gift certificate.

5. A *reinforcement plan* determines
   - desired results for selected task(s);
   - specific behaviors to obtain results;
   - measures of results in terms of quality, quantity, costs, or timeliness;
   - feedback about past behavior that can help an individual change (specific, sincere, immediate, and personal);
   - positive reinforcement plan for pinpointed results and behaviors.

The behaviors monitored will achieve the desired results if the corporation provides good procedures, proper equipment, and resources. If desired results are not obtained (the metrics) then the organization can easily find out whether the root cause is the manufacturing technicians (not following the standards) or the organization (high number of workarounds). See response to point 8 below.

6. *Feedback mechanism:*
   - Immediate feedback is desired but needs to be defined.
   - It must be easy to monitor and maintain.
   - It must be specific.

The program is easy to administer and feedback is immediate if not following standards has a critical processing impact, or, if not following standards has no immediate impact, slightly delayed in order not to interfere with manufacturing processing. The feedback from should be one page and relatively easy to use.

7. *Implementation plan:*
   - How will the plan be rolled out and when will it be implemented?

A rollout plan should be specific and is dependent on the organization structure, staffing, schedules, etc. A plan is not included in this example.

8. *Metrics* to determine the effectiveness:
   (a) Stretch goal: Achieve a 10 percent reduction in documentation errors per million opportunities per month.
   (b) Achieve fewer than 100 documentation errors per million opportunities per month.

If the staff follows all steps in a manufacturing procedure, this ensures consistency in terms of safety, quality, and purity of the product, and ability to identify the product.

Example: Upon implementation of this program, the observers find that the manufacturing technicians are not meeting their goals due to inadequate batch records. Upon investigation, they find that three manufacturing records contain all the documentation errors. The organization can go back to the observation program and look at observation data gathered during that time and determine that the manufacturing technicians were meeting the documentation standards of the program by contacting supervision, but mistakes were still being made. With the added information from the observation data, the organization now knows that it is the record that is contributing to the mistakes and not the manufacturing technicians' documentation practices during processing. Without the observation program data, this would not be apparent.

In a manufacturing setting, document authors often contend that their records are not contributing to mistakes, but that technicians are making the mistakes and have sole ownership of their errors. This is typical human nature. At the same time, manufacturing technicians will indicate that the records are contributing to or are the source of their errors. Having observation program data provides the organization with the necessary information to objectively know what is contributing to the errors (people, process, documents, etc.) so the organization can prioritize resources and projects to properly resolve errors.

One of three results would be an outcome of this example:

1. Behaviors need correcting, preventing the metrics from being met.
2. Batch records need correcting, preventing metrics from being met.
3. Job observation program needs correcting: either
   - metrics are not being met, but behavioral standards are being met, and the corporation is supplying all of the necessary tools for success (adequate documents, equipment, etc.), or
   - metrics are being met, and behavioral standards are not being met (consider modifying behavioral standards).

The following two items are not specifically in this example:

9. Duration of an evaluation of the GMP documentation observation program
10. Changes to the GMP documentation observation program based on evaluation

Appendix C:
Test

1. Which sections of the CFR describe the regulatory requirements for verification and independent verification?
2. What is the purpose of a Job Observation Program?
3. Name the 8 items needed for a Job Observation Program.
4. List the 10 attributes of a Job Observation Program that should be addressed.
5. Why is procedural adherence in drug manufacturing important?

Test and Test Answers

1. What sections of the CFR describe the regulatory requirements for verification and independent verification?
   Answer: 21 CFR Part 211.103 and 21CFR Part 211.101
2. What is the purpose of a job observation program?
   Answer: To modify behaviors
3. Name the eight items needed for a job observation program.
   Answer:
   1. *Expectations for the group of people who will be required to carry out the actions*
   2. *Actions that satisfy a specific definition as to what is acceptable and what is unacceptable*
   3. *Job observers*
   4. *Job observation/inspections activities (points to monitor)*

5. *Reinforcement plan*

6. *Immediate feedback through a defined feedback mechanism*

7. *Implementation plan defining timing and implementation method*

8. *Metrics to determine effectiveness that are objective and measurable*

4. List the 10 attributes of a job observation program that should be addressed

   Answer:

   1. *Setting where the observations take place*

   2. *Specific behaviors to observe*

   3. *Development of the documentation process to use to track the observations and effectiveness of the program*

   4. *Feedback mechanism the program will use, including the timing of feedback*

   5. *Specific criteria used during the observations*

   6. *The reward or punitive system to implement with the program*

   7. *Method to measure program effectiveness, to include metrics, what is being measured, goals*

   8. *Implementation plan for the program*

   9. *Duration and evaluation of the program*

   10. *Changes to the program based on evaluation*

5. Why is procedural adherence in drug manufacturing important?

   Answer: Adherence to established procedures in a regulated environment is essential and fundamental to maintaining regulatory compliance and quality control of a manufacturing process.

Resources

Code of Federal Regulations (2006). Available at: http://www.gpoacces.gov/cfr/index.html, sec. 211.103 & sec.211.101 (last accessed 24 September 2008).

Food and Drug Administration (FDA): http://www.fda.gov

Human Factors and Ergonomics Society (HFES): http//wwwhfes.org/web/Default.aspx

Subject Index

a

abbreviated new drug application
    (ANDA) 59, 106–107
absolute novelty requirement 86
absorption profile 59
abstract concepts, patentability 83
abuse liability 59
academia *see* university
academic skills 4
acquired immune deficiency syndrome 19
actions, job observation program 183
ADME 59
adoption, of new technological
    advances 145
adverse reaction 59
AIDS 19
alliance dynamics 5
allowance, of patent application 104
Alzheimer's disease, priority ranking 48
American Type Culture Collection
    (ATCC) 91
amortized cost, of developing a new
    drug 24
analyze (process step), Six Sigma
    principle 147
ANDA *see* abbreviated new drug application
    angel investors 73
assay type, and marker analyte 37
assay validity 37
assessing opportunity, entrepreneur-
    ship 68–70
assignment, of a patent 111
ATCC *see* American Type Culture Collection
Atomic Energy Act 83

b

barring, of a patent 87
BCR-ABL fusion protein 29
BCR-ABL gene 29

behavior standards
– cooking 185
– GMP documentation 190
best mode, patentability requirement 92
biofilms
– effect on quality 162, 169–172
biological license application (BLA) 59
biomarkers
– and diagnostic product development 22
– applications 26–27
– drug development acceleration 30–32
biotechnological product, development
    costs 24
biotechnology industries
– career in 62–63
– skills in high demand 4
biotechnology inventions 84, 91, 101, 103
biotechnology related companies, early-
    stage 73
biotechnology venture, establishing 61–77
BLA *see* biological license application
board of directors, start-up company 71
breakthrough idea 65–66
breast cancer
– need for chemotherapy 20
– therapy 27–28
brown paper sessions 146
Budapest Treaty 91
business, financing of 72–73
business development, case study 7
business model, description in business
    plan 70
business plan
– components 69
– development 62
– writing a 68–70

Industry Immersion Learning. Real-Life Industry Case-Studies in Biotechnology and Business
L. Borbye, M. Stocum, A. Woodall, C. Pearce, E. Sale, W. Barrett, L. Clontz, A. Peterson, J. Shaeffer
Copyright © 2009 WILEY-VCH Verlag GmbH & Co. KGaA, Weinheim
ISBN: 978-3-527-32408-8